BASICS OF
QUANTUM MECHANICS

BASICS
OF
QUANTUM MECHANICS

Pankaj Jha

ANMOL PUBLICATIONS PVT. LTD.
NEW DELHI - 110 002 (INDIA)

ANMOL PUBLICATIONS PVT. LTD.
H.O.: 4374/4B, Ansari Road, Daryaganj,
New Delhi-110 002 (India)
Ph.: 23278000, 23261597
B.O.: No. 1015, Ist Main Road, BSK IIIrd Stage
IIIrd Phase, IIIrd Block,
Bangalore - 560 085 (India)
Visit us at: www.anmolpublications.com

Basics of Quantum Mechanics

First Published, 2008

ISBN 978-81-261-3565-3

PRINTED IN INDIA

Printed at Mehra Offset Press, Delhi.

Contents

Preface

By now it is hard to imagine an instructor who has not heard the call to "teach with technology," as it has resounded through educational institutions and government agencies alike over the past several years. It is, however, easier to imagine an instructor of modern physics and quantum mechanics who has not heard of the current research into the teaching and learning by author. Despite this work, which focuses on improving the conceptual understanding of students, the teaching of quantum mechanics has remained relatively unchanged since its inception. Students, therefore, often see quantum physics in terms of misleading (such as the *convention* of drawing wave functions on potential energy diagrams) or incomplete visualizations, and as one dimensional and time independent (because of the focus on energy eigenstates in one dimension), and devoid of almost any connection with classical physics. These depictions short change quantum physics. Quantum physics is a far richer topic when non-trivial time evolution, multiple dimensions, classical-quantum connections, and research-based topics are discussed.

Physlet Quantum Physics is an interactive text with over 200 ready-to-run interactive exercises which use over 250 carefully-designed computer simulations for the teaching of quantum physics.[1] This material uses a standard easy-to-understand interface designed with a sound use of pedagogy in mind. The aim of *Physlet Quantum Physics* is to provide a resource for the teaching of quantum physics that enhances student learning through interactive engagement and visualization. At the same time, *Physlet Quantum Physics* is a resource flexible enough to be adapted to a variety of pedagogical strategies and local environments, covers a wide variety of topics, and is informed by current educational, experimental, and theoretical research.

Author

Chapter 1

Introduction

WHAT WAS WRONG WITH CLASSICAL MECHANICS?

Basically, classical statistical mechanics wasn't making sense...Maxwell and Boltzmann evolved the equipartition theorem: a physical system can have many states (gas with particles having different velocities, or springs in different states of compression). At nonzero temperature, energy will flow around in the system, it will constantly move from one state to another. So, what is the probability that at any instant it is in a particular state with energy E? M and B proves it was proportional to $e^{-E/kT}$. This proportionality factor is also correct for any subsystem of the system: for example a single molecule.

Notice this means if a system is a set of oscillators, different masses on different strength springs, for example, then in thermal equilibrium *each oscillator has on average the same energy as all the others*. For three-dimensional oscillators in thermal equilibrium, the average energy of each oscillator is $3kT$, were k is Boltzmann's constant.

BLACK BODY RADIATION

Now put this together with Maxwell's discovery that light is an electromagnetic wave: inside a hot oven, Maxwell's equations can be solved yielding standing wave solutions, and the set of different wavelength allowed standing waves amount to an infinite series of oscillators, with no upper limit on the frequencies on going far into the ultraviolet. Therefore, from the classical equipartition theorem, an oven at thermal equilibrium at a definite temperature should contain an infinite amount of energy of order kT in each of an infinite number of modes and if you let radiation out through a tiny hole in the side, you should see radiation of all frequencies.

This is not, of course, what is observed: as an oven is warmed, it emits infrared, then red, then yellow light, etc. This means that the higher frequency oscillators (blue, etc.) are in fact *not* excited at low temperatures: equipartition isn't true.

Planck showed that the experimentally observed intensity/frequency curve was exactly reproduced if it was assumed that the radiation was *quantized*: light of frequency f could only be emitted in quanta now photons having energy hf, h being Planck's constant. This was the beginning of quantum mechanics.

THE PHOTOELECTRIC EFFECT

Einstein showed the same quantization of electromagnetic radiation explained the photoelectric effect: a photon of energy hf knocks an electron out of a metal, it takes a certain work W to get it out, the rest of the photon energy goes to the kinetic energy of the electron, for the fastest electrons emitted. Plotting the maximum electron kinetic energy as a function of incident light frequency confirms the hypothesis, giving the same value for h as that needed to explain radiation from an oven.

THE BOHR ATOM

Bohr put together this quantization of light energy with Rutherford's discovery that the atom had a nucleus, with electrons somehow orbiting around it: for the hydrogen atom, light emitted when the atom is thermally excited has a particular pattern, the observed emitted wavelengths are given by $\frac{1}{\lambda} = R_H\left(\frac{1}{4} - \frac{1}{n^2}\right)$, $n = 3, 4, 5$. (R_H is now called the Rydberg constant.) Bohr realised these were photons having energy equal to the *energy difference between two allowed orbits* of the electron circling the nucleus (the proton), $E_n - E_m = hf$, leading to the conclusion that the allowed levels must be $E_n = -hcR_H/n^2$. How could the quantum hf restricting allowed radiation energies also restrict the allowed electron orbits? Bohr realised there must be a connection because h has the dimensions of angular momentum!

What if the electron were only allowed to be in circular orbits of angular momentum nKh, with n an integer? Bohr did the math for orbits under an inverse square law, and found that the observed spectra were in fact correctly accounted for by taking $K = 1/2\pi$. But then he realised he didn't even need the experimental results to find K: quantum mechanics *must* agree with classical mechanics in the regime where we know experimentally that classical mechanics (including Maxwell's equations) is correct, that is, for systems of macroscopic size. Consider a negative charge orbiting around a fixed positive charge at a radius of 10cm., the charges being such that the speed is of order meters per second. Then from classical E and M, the charge will radiate at the orbital frequency. Now imagine this is actually a hydrogen atom, in a perfect vacuum, in a high state of excitation. It must be radiating at this same frequency. But Bohr's theory can't just be right for small orbits, so the radiation must satisfy $E_n - E_m = hf$. The spacing between adjacent levels will vary slowly for these large orbits, so h times the orbital frequency must be the energy difference between adjacent levels. Now, that energy difference depends on the allowed angular momentum step between the adjacent levels: that is, on K. Reconciling these two expressions for the radiation frequency gives $K = 1/2\pi$.

This *classical limit argument, then, predicts the Rydberg constant* in terms of already known quantities:

$$R_H = \left(\frac{1}{4\pi\varepsilon_0}\right)^2 \cdot \frac{2\pi^2 me^4}{ch^3}$$

What's right about the Bohr atom?

1. It gives the Balmer series spectra.
2. The first orbit size is close to the observed size of the atom: and remember there are no adjustable parameters, the classical limit argument determines the spectra and the size.

What's *wrong* with the Bohr atom?

1. No explanation for why angular momentum should be quantized. This was solved by de Broglie a little later.
2. Why don't the circling electrons radiate, as predicted classically?

   Well, the fact that radiation is quantized means the classical picture of an accelerating charge smoothly emitting radiation cannot work if the energies involved are of order h times the frequencies involved.
3. The lowest state has nonzero angular momentum.

   This is a defect of the model, corrected in the truly quantum model (Schrodinger's equation).
4. In an inverse square field, orbits are in general elliptical.

This was at first a puzzle: why should there be only circular orbits allowed? In fact, the model does allow elliptical orbits, and they don't show up in the Balmer series because, as proved by Sommerfeld, if the allowed elliptical orbits have the same allowed angular momenta as Bohr's orbits, they have the same set of energies. This is a special property of the inverse square force.

DE BROGLIE WAVES

The first explanation of why only certain angular momenta are allowed for the circling electron was given by de Broglie: just as photons act like particles (definite energy and momentum), but undoubtedly are wave like, being light, so particles like electrons perhaps have wave like properties. For photons, the relationship between wavelength and momentum is $p = h/\lambda$. Assuming this is also true of electrons, and that the allowed circular orbits are standing waves, Bohr's angular momentum quantization follows.

SCHRODINGER'S WAVE EQUATION

De Broglie's idea was clearly on the right track but waves in space are three-dimensional, thinking of the circular orbit as a string under tension can't be right, even if the answer is.

Photon waves (electromagnetic waves) obey the equation. $\nabla^2 \vec{E} - \frac{1}{c^2}\frac{\partial^2 \vec{E}}{\partial t^2} = 0$

A solution of definite momentum is the plane wave

$$\left(\frac{\partial^2}{\partial x^2} - \frac{1}{c^2}\frac{\partial^2}{\partial t^2}\right)\vec{E}_0 e^{i(kx-\omega t)} = \left(k^2 - \frac{\omega^2}{c^2}\right)\vec{E}_0 e^{i(kx-\omega t)} = 0$$

Notice that the last equality is essentially just $\omega = ck$, where for a plane wave solution the energy and momentum of the photon are translated into differential operators with respect to time and space respectively, to give a differential equation for the wave.

Schrodinger's wave equation is equivalently taking the (nonrelativistic) energy-momentum relation $E = p^2/2m$ and using the same recipe to translate it into a differential equation:

$$i\hbar\frac{\partial \psi(x,t)}{\partial t} = -\frac{\hbar^2}{2m}\frac{\partial^2 \psi(x,t)}{\partial x^2}$$

Making the natural extension to three dimensions, and assuming we can add a potential term in the most naïve way possible, that is, going from $E = p^2/2m$ to $E = p^2/2m + V(x, y, z)$, we get

$$i\hbar\frac{\partial \psi(x,y,z,t)}{\partial t} = -\frac{\hbar^2}{2m}\nabla^2\psi(x,y,z,t) + V(x,y,z)\psi(x,y,z,t).$$

This is the equation Schrodinger wrote down and solved, the solutions gave the same set of energies as the Bohr model, but now the ground state had zero angular momentum.

A CONSERVED CURRENT

Schrodinger also showed that a conserved current could be defined in terms of the wave function ψ.

$$\frac{\partial \rho}{\partial t} + \text{div}\,\vec{j} = 0$$

$$\text{where } \rho = \psi^* \psi = |\psi|^2$$

$$\text{and } \vec{j} = \frac{\hbar}{2mi}(\psi^* \vec{\nabla} \psi - \psi \vec{\nabla} \psi^*).$$

Schrodinger's interpretation of his equation was that the electron was simply a wave, not a particle, and this was the wave intensity. But thinking of electromagnetic waves in this way gave no clue to the quantum photon behaviour this couldn't be the whole story.

INTERPRETING THE WAVE FUNCTION

The correct interpretation of the wave function follows from analogy to the electromagnetic case. Let's review that briefly. The basic example is the two-slit diffraction pattern, as built up by sending through one photon at a time, to a bank of photon detectors. The pattern gradually emerges: solve the wave equation, then the predicted local energy density (proportional to $|E(x, y, z, t)|^2\, dxdydz$) gives the probability of one photon going through the system landing at that spot.

Born suggested that similarly $|\psi|^2$ at any point was proportional to the probability of detecting the electron at that point. This has turned out to be correct.

LOCALIZING THE ELECTRON

Despite its wavelike properties, we know that an electron can behave like a particle: specifically, it can move as a fairly localized entity from one place to another. What's the wave representation of that? It's called a wave packet: a localized wave excitation. To see how this can come about, first remember that the Schrodinger equation is a linear equation, the sum of any two or more solutions is itself a solution. If we add together two plane waves close in wavelength, we get beats, which can be regarded as a string of wave packets. To get a single wave packet, we must add together a continuous range of wavelengths.

The standard example is the Gaussian wave packet, $\psi(x,t=0) = Ae^{ik_0x}e^{-x^2/2a^2}$ where $p_0 = \hbar k_0$.

Using the standard result

$$\int_{-\infty}^{+\infty} e^{-ax^2}\,dx = \sqrt{\frac{\pi}{a}}$$

we find $|A^2| = \left(\pi\Delta^2\right)^{-1/2}$ so

$$\psi(x,t=0) = \frac{1}{\left(\pi\Delta^2\right)^{1/4}} e^{ik_0x_e^{-x^2/2\mu 2}}$$

But how do we construct this particular wavepacket by superposing plane waves? That is to say, we need a representation of the form:

$$\psi(x) = \int_{-\infty}^{+\infty} \frac{dk}{2\pi} e^{ikx}\phi(k)$$

The function $\phi(k)$ represents the weighting of plane waves in the neighbourhood of wavenumber k. This is a particular example of a *Fourier transform*. Note that if $\phi(k)$ is a bounded function, any particular k value gives a vanishingly small contribution, the plane-wave contribution to $\psi(x)$ from a range dk is $\phi(k)dk/2\pi$ In fact, $\phi(k)$ is given in terms of $\psi(x)$ by

$$\phi(k) = \int_{-\infty}^{+\infty} dxe^{-ikx}\psi(x)$$

It is perhaps worth mentioning at this point that this can be understood *qualitatively* by observing that the plane wave prefactor e^{-ikx} will interfere destructively with all plane wave components of $\psi(x)$ except that of wavenumber k, where it may at first appear that the contribution is infinite, but recall that as stated above, any particular k component has a vanishingly small weight.

In the present case, the above handwaving argument is unnecessary, because both the integrals can be carried out exactly, using the standard result:

$$\int_{-\infty}^{\infty} e^{-ax^2+bx}dx = e^{b^2/4a}\sqrt{\frac{\pi}{a}}$$

giving

$$\phi(k) = (4\pi\Delta^2)^{\frac{1}{4}} e^{-\Delta^2(k-k_0)^2/2}$$

THE UNCERTAINTY PRINCIPLE

Note that the spreads in x-space and p-space are inversely related: Δx is of order Δ, $\Delta p = \hbar\Delta k \sim \hbar/\Delta$. This is of course the Uncertainty Principle, localization in x-space requires a large spread in contributing momentum states.

It's worth reviewing the undergraduate exercises on applications of the uncertainty principle. The help sharpen one's appreciation of the wave/particle nature of quantum objects.

There's a limit to how well the position of an electron can be determined: it is detected by bouncing a photon off of it, and the photon wavelength sets the limit on Δx. But if the photon has enough energy to create an electron-positron pair out of the vacuum, you can't be sure which electron you're seeing. This limits $\Delta x \sim \hbar/mc$ at best. (This is called the Compton wavelength, written λ_C – it appears in Compton scattering.) How much smaller that a hydrogen atom ground state wave function is this? $\lambda_c/a_0 = e^2/\hbar c(CGS)$ =

$e^2/4\pi\varepsilon_0\hbar c(SI)=1/137$, known as the fine structure constant. This is also the ratio of the electron speed in the first Bohr orbit to the speed of light, and so is an indication of the importance of relativistic corrections to energies of electron states; these differences in electron orbit energies for circular and elliptical states having the same energy when calculated nonrelativistically lead to fine structure in the atomic spectra.

Chapter 2

History of Quantum Mechanics

WHY DO WE NEED QUANTUM MECHANICS?

Just over 100 years ago, in the 1890's, physics looked in pretty good shape. The beautiful mathematical development of Newton's mechanics, coupled with increasingly sophisticated technology, predicted the movements of the solar system to incredible accuracy, apart from a tiny discrepancy in the orbit of Mercury. It had been less than a hundred years since it was realised that an electric current could exert a force on a magnet, but that discovery had led to power stations, electric trains and a network of telegraph wires across land and under the oceans. It had also been only a hundred years since it had been established that light was a wave, and only forty years since Maxwell's realization that the waves in a light signal were electric and magnetic fields, satisfying a wave equation he was able to derive purely by considering electric and magnetic field phenomena. In particular, he was able to predict the speed of light by measuring the electrostatic attractive forces between charges and the magnetic forces between currents.

At about the same time, in the 1860's, Maxwell and Boltzmann gave a brilliant account of the properties of gases by assuming that they were made up of weakly interacting molecules flying about in a container, bouncing off the sides, with a statistical distribution of energies so that the probability of a molecule having energy E was proportional to $e^{-E/kT}$, k being a universal constant known as Boltzmann's constant.

Boltzmann generalized this result from a box of gas to any system. For example, a solid can be envisioned classically as a lattice of balls (the atoms) connected by springs, which can sustain oscillations in many different ways, each such mode can be thought of as a simple harmonic oscillator, with reasonable approximations concerning the properties of the springs, etc. Boltzmann's work leads to the conclusion that each such mode of oscillation, or degree of freedom, would at temperature T have average energy kT, made up of $\frac{1}{2}kT$ potential energy, $\frac{1}{2}kT$ kinetic energy. Notice that this average energy is independent of the strength of the springs, or the masses! *All* modes of vibration, which will vibrate at very different rates, contain the same energy at the same temperature. This equal sharing is called *the Equipartition of Energy*. It is not difficult to check this for a one-dimensional classical harmonic oscillator, averaging the energy by integrating over all displacements and momenta (independently) with the weighting factor $e^{-E/kT}$ (which of course needs to be normalized). The result doesn't depend on the spring constant or the mass. Boltzmann's result gave an excellent account of the specific heats of a wide range of materials over a wide temperature range, but there *were* some exceptions, for example hydrogen gas at low temperatures, and even solids at low enough temperatures. Still, it was generally felt these problems could be handled within the existing framework, just as the slightly odd behaviour of Mercury was likely caused by a small planet, named Vulcan, closer to the sun, and so very hard to observe.

BLACK BODY RADIATION

But there was one problem that was hard to get a grip on, an apparently blatant violation of the equipartition of energy. Consider an oven with a small hole in the door, through which the radiation inside is observed. This oven can be heated until it's white hot. The radiation inside is infrared at low temperatures, becoming visible light as the temperature increases. So, the oven's full of electromagnetic waves, satisfying Maxwell's wave equation, with boundary conditions at the walls of the oven, the electric field has to be essentially zero there, because the walls conduct currents. Of course, the radiation originates in oscillating charges in the walls, using the same analysis of Maxwell's equations that gives the radiation form an antenna. Anyway, there is a set of standing wave modes of electromagnetic vibrations inside the oven, just a three-dimensional version of the series of allowed standing wave modes of vibration of a string fixed at both ends. So, we should be able to find the energy density of these waves using the same ideas that worked pretty well for the specific heats of solids and gases, that is to say, assume there's kT of energy in each mode of vibration. (This is ½kT of kinetic energy, ½kT of potential energy for each independent direction of vibration.)

But this leads to disaster. The problem is that there are infinitely many modes of vibration of the electromagnetic field in an oven. There is no upper limit to the number of wiggles the wave can have between the walls. So, if we take kT in each mode, we deduce that the oven contains an infinite amount of energy, and radiates an infinite amount through our small hole. This analysis gives no clue as to why the colour we see changes with temperature. Evidently, equipartition of energy isn't working in this case. There's only a finite amount of energy in the oven and at low temperatures there's no energy at all in the modes corresponding to visible light, although that changes as things get hotter.

In the 1890's, German experimentalists measured the energy density as a function of wavelength to great precision, it's called the black body radiation spectrum. A theorist, Planck, found a mathematical formula that fitted this curve exactly,

$$R_T(f)df = \frac{8\pi Vhf^3 df / c^3}{e^{hf/kT} - 1}$$

He did not at first have any theoretical justification for this formula, but it was a very accurate fit to some very precise experiments for a suitable value of the constant h, which we discuss in a moment.

Factoring out the number of modes of oscillation in the frequency range df, Planck's formula gives the average energy per mode to be

$$\frac{hf}{e^{hf/kT}-1}$$

For low frequencies, $hf \ll kT$, this correctly gives kT per mode.

But, for higher frequencies it's clear that the oscillators are not getting their "fair share" kT of energy. Somehow, the oscillating charges in the walls are not radiating so much energy at the high frequencies. The only way Planck could derive the formula theoretically was by making a weird assumption: he assumed that the oscillating charges in the walls could *not* just radiate energy continuously, as Maxwell's equations would predict (and as was known to be true for ordinary antennas) but were only allowed to radiate energy in chunks he called *quanta*. Furthermore, the amount of energy in one quantum depended on the frequency of the oscillation, in fact linearly: for frequency f, the quantum has energy hf, where h is the constant introduced into the formula above, now known as Planck's constant. It follows that the oscillators themselves could only be oscillating with energies that form a ladder with steps hf apart, above some lowest energy which would be their energy at absolute zero temperature.

The formula follows if we assume the oscillating field component in the oven having frequency f can only have a whole number of quanta of energy, that is to say, its energy must be one of: $0, hf, 2hf, 3hf, \ldots$ The relative probability of it having energy E is $e^{-E/kT}$, then its relative probabilities of having energy $0, hf, 2hf, \ldots$ are in the ratio $1: e^{-hf/kT} : e^{-2hf/kT}$, etc.

The actual probabilities are given by dividing these relative probabilities by the sum of all of them. They clearly are the terms of a geometric series, so their sum is just $1/(1 - e^{-hf/kT})$. So, to find the average energy in the oscillator, we take the possible energies $0, hf, 2hf, 3hf, \ldots$ and weight each of them with their probability of occurring, that is, we must find

$$0.1 + hf.e^{-hf/kT} + 2hf.e^{-2hf/kT} + ...,$$

and divide the sum by $1/(1 - e^{-hf/kT})$.

So, Planck's quantum assumption explains the observed black body radiation curve. It also gives a qualitative explanation of the change in colour of the radiated light as the temperature is increased. The oscillators in the walls derive their energy from the heat vibrations of neighbouring molecules: typically, such a vibration has energy of order kT, with probabilities of more energy going down as $e^{-E/kT}$. This means that if the potentially radiating oscillator can only absorb energies in quanta hf, if $kT \ll hf$, it will be very unlikely to absorb *any* energy, and therefore very unlikely to radiate. In the three-dimensional oven, the number of standing wave oscillations in a small frequency range Δf increases with f as f^2, so we find that the maximum radiation intensity occurs at a frequency f such that hf is of order kT. Therefore, as the temperature increases, the frequency at which the most intense radiation occurs increases, and hence the colour moves from red to blue.

THE PHOTOELECTRIC EFFECT

If light shines on certain metals, electrons are emitted. This is the photoelectric effect. If the metal is in air, the electrons bounce off air molecules and are almost certainly rapidly reabsorbed, but if the metal surface is in a vacuum, the electrons can fly away, and in a vacuum tube they can be collected by another piece of metal, and light can cause a current to flow, the origin of the photoelectric cell. In 1902, Lenard studied how the energy of the emitted photoelectrons varied with the intensity of the light. He used a carbon arc light, and could increase the intensity a thousand-fold. The ejected electrons hit another metal plate, the collector, which was connected to the cathode by a wire with a sensitive ammeter, to measure the current produced by the illumination. To measure the energy of the ejected electrons, Lenard charged the collector plate negatively, to repel the electrons coming towards it. Thus, only electrons ejected with enough kinetic energy to get up this potential hill would contribute to the current. Lenard discovered that there was a well-defined minimum voltage that stopped any electrons getting through, we'll call it V_{stop}. To his surprise, he found that V_{stop} did not depend at all on the intensity of the light! Doubling the light intensity doubled the number of electrons emitted, but did not affect the energies of the emitted electrons. He also discovered, by using light of different colours, that the maximum electron energy did increase as the frequency of the incident light increased.

EINSTEIN SUGGESTS AN EXPLANATION

In 1905 Einstein gave a very simple interpretation of Lenard's results. He just assumed that the incoming radiation should be thought of as quanta of frequency hf, with f the frequency. In photoemission, one such quantum is absorbed by one electron. If the electron is some distance into the material of the cathode, some energy will be lost as it moves towards the surface. There will always be some electrostatic cost as the electron leaves the surface, this is usually called the work function, W. The most energetic electrons emitted will be those very close to the surface, and they will leave the cathode with kinetic energy

$$E = hf - W.$$

On cranking up the negative voltage on the collector plate until the current just stops, that is, to V_{stop}, the highest kinetic energy electrons must have had energy eV_{stop} on leaving the cathode. Thus,

$$eV_{stop} = hf - W$$

Thus Einstein's theory makes a very definite quantitative prediction: if the frequency of the incident light is varied, and V_{stop} plotted as a function of frequency, the slope of the line should be h/e. It is also clear that there is a minimum light frequency for a given metal, that for which the quantum of energy is equal to the work function. Light below that frequency, no matter how bright, will not cause photoemission.

MILLIKAN'S ATTEMPTS TO DISPROVE EINSTEIN'S THEORY

If we accept Einstein's theory, then, this is a *completely different* way to measure Planck's constant. The American experimental physicist Robert Millikan, who did not accept Einstein's theory, which he saw as an attack on the wave theory of light, worked for ten years, until 1916, on the photoelectric effect, to disprove Einstein's theory. He even devised techniques for scraping clean the metal surfaces inside the vacuum tube. For all his efforts he found disappointing results: Einstein's theory, measuring Planck's constant to within 0.5per cent by this method. One consolation was that he did get a Nobel prize for this series of experiments.

The point to be emphasized is that the same value for Planck's constant, 6.6×10^{-34} Joule.sec, emerges from two completely different experiments: the measurement of black body radiation, and measuring energies of emitted electrons in the photoelectric effect. This is clearly a general property of electromagnetic radiation, and is confirmed by many later experiments, for example Compton scattering, in which light scatters off electrons. By measuring the energy change and momentum change of the electron, it is found that a single quantum of light was scattered.

THE NATURE OF LIGHT

It is firmly established experimentally that the propagation of light is well described by a wave equation, which in fact is not difficult to derive from Maxwell's equations:

$$\nabla^2\vec{E}-\frac{1}{c^2}\frac{\partial^2\vec{E}}{\partial t^2}=0$$

For a plane wave moving in the x-direction this reduces to

$$\frac{\partial^2\vec{E}}{\partial x^2}-\frac{1}{c^2}\frac{\partial^2\vec{E}}{\partial t^2}=0$$

The monochromatic solution to this wave equation has the form

$$\vec{E}(x,t) = \vec{E}_0 e^{i(kx-\omega t)}$$

(Another possible solution is proportional to $\cos(kx-\omega t)$.We shall find that the exponential form, although a complex number, proves more convenient. The physical electric field can be taken to be the real part of the exponential for the classical case.)

Applying the wave equation differential operator to our plane wave solution

$$\left(\frac{\partial^2}{\partial x^2} - \frac{1}{c^2}\frac{\partial^2}{\partial t^2}\right)\vec{E}_0 e^{i(kx-\omega t)} = \left(k^2 - \frac{\omega^2}{c^2}\right)\vec{E}_0 e^{i(kx-\omega t)} = 0$$

If the plane wave is a solution to the wave equation, this must be true for all x and t, so we must have

$$\omega = ck$$

Solving this equation for boundary conditions like an antenna can be quite challenging, but all we need consider at the moment is some illustration of diffraction. We take the case of a double slit experiment: if a plane wave encounters a barrier with two equal narrow parallel slit openings, the transmitted wave reaching a screen some distance further on will show a series of bright and dark stripes parallel to the slits.This pattern can be quantitatively accounted for. The two slits transmit radiation in phase with each other. At each point on the screen, the electric field vector from slit 1 must be added to the electric field vector from slit 2. At a point on the screen equidistant from the two slits, the electric field vectors will be equal. Moving away from that point in a direction perpendicular to the slits we will reach a point where the field from one slit is exactly out of phase with the field from the other slit the screen will be dark.

In fact, the intensity if the light at any point on the screen is proportional to $|E_0|^2$.

Now consider what happens as we make the light dimmer and dimmer. How easy is it to see this diffraction pattern? Eventually we need to soup up our detection apparatus. We replace our screen and visual inspection with a series of photodetectors. Experimentally, we find that, just as in the photoelectric effect, our detectors will only detect quanta, just as if the light were made up of particles, photons. Suppose now we dim the light so that our photodetectors only detect one photon per minute coming through the slits. If we record where each photon lands, and build up a picture, we find the very same pattern of light and dark stripes that we saw with bright light. In other words, if we send through one photon, we can't predict where it will land, but if we send through a thousand, we will begin to discern the stripes. The best we can do for one photon is to say it will more probably land where the solution to Maxwell's wave equation gives a large $|E_0|^2$. That is to say, $|E_0(x)|^2$ is proportional to the probability of the photon being at x.

But this means each photon must have gone through both slits! The probability distribution for a single photon is given by the stripes, and the distance between the stripes depends on the distance between the slits. The photon, therefore, knows about both slits. So the bottom line is: to find where one photon will be, solve the wave equation to find the electric field everywhere on the screen. The probability of the photon landing at any particular point is proportional to $|E_0|^2$ at that point.

THE NATURE OF MATTER

By the 1890's and early 1900's, most scientists believed in the existence of atoms. Not all the distinguished German chemist Ostwald didn't, for example. But nobody had a clear picture of even a hydrogen atom. The electron had just been discovered, and it was believed that the hydrogen atom had a single electron. It was suggested that maybe the electron went in circles around a central charge, but nobody believed that because Maxwell had established that accelerating charges radiate, so it was assumed that a circling electron would rapidly loose energy, spiral in to the center, and the atom would collapse. Instead, it was thought, the hydrogen atom (which was of course electrically neutral) was a ball of positively charged jelly with an electron inside, which would oscillate when heated, and emit radiation. Rough calculations, based on the accepted size of the atom, suggested that the radiation would be in the visible range, but no-one could remotely reproduce the known spectrum of hydrogen.

The big breakthrough came in 1909, when Rutherford tried to map the distribution of positive charge in a heavy atom (gold) by scattering alpha particles from it. To his amazement, he found the positive charge was all concentrated in a tiny nucleus, with a radius of order one ten-thousandth that of the atom. This meant that after all the electrons must be going in some kind of planetary orbits, and the Maxwell's equations prediction of radiation didn't apply, just as it didn't always apply in black body radiation.

THE BOHR ATOM

The Danish theorist Niels Bohr was visiting Manchester at the time Rutherford did this experiment, and Bohr decided that there must be certain allowed sets of electron orbits in the atom where the classical acceleration radiation didn't occur: he called them "stationary states". The lowest energy stationary state would be the ground state of the atom, the others would eventually go to that state by emitting photons corresponding to energy differences between states. But Bohr was of the opinion that looking at the very complex spectra emitted by heated atoms would never be helpful he remarked that it would be like trying to understand fundamental biology by studying the colours of butterfly wings. He changed his mind in February 1913, when a casual conversation with the spectroscopist H. R. Hansen revealed that one pattern *had* been discerned in the apparent chaos of spectral lines. In particular, Hansen (a colleague and former classmate of Bohr) showed him Balmer's formula for hydrogen. Balmer was a math and Latin teacher at a girls' school in Switzerland, and had found his formula in the 1880's.

Balmer's formula is: $\frac{1}{\lambda}=R_H\left(\frac{1}{4}-\frac{1}{n^2}\right)$ for the sequence of wavelengths of light emitted, with n = 3, 4, 5, 6 being in the visible, the lines used by Balmer in finding the formula. Hansen would doubtless have informed Bohr that the 1/4 could be replaced by $1/m^2$, with m another integer. The constant appearing on the right hand side is called the *Rydberg constant*, R_H = 109,737 cm^{-1}. Bohr said later: "As soon as I saw Balmer's formula, the whole thing was immediately clear to me." What he saw was that the set of allowed *frequencies* (proportional to inverse wavelengths) emitted by the hydrogen atom could all be expressed as *differences*. This immediately suggested to him a generalization of his idea of a "stationary state" lowest energy level, in which the electron did not radiate. There must be a *whole sequence* of these stationary states, with radiation only taking place as the atom jumps from one to another of lower energy, emitting a single quantum of frequency f such that

$hf = E_n - E_m$, the difference between the energies of the two states.

Evidently, from the Balmer formula and its extension to general integers m, n, these allowed non-radiating orbits, the stationary states, could be labeled 1, 2, 3, ... , n, ... and had energies $E_n=-hcR_H/n^2$ using $\lambda f=c$ and the Balmer equation above. The energies are of course negative, because these are bound states, and we take the zero of energy to be where the two particles are at rest infinitely far apart.Bohr was very familiar with the dynamics of simple circular orbits in an inverse square field. He knew that if the energy of the orbit was - hcR_H/n^2, that meant the kinetic energy of the electron, $\frac{1}{2}mv^2 = hcR_H/n^2$, and the potential energy would be

$$-\frac{1}{4\pi\varepsilon_0}\cdot\frac{e^2}{r_n}=-\frac{2hcR_H}{n^2}$$

It immediately follows that the *radius* of the nth orbit is proportional to n^2, and the *speed* in that orbit is proportional to $1/n$.

It then follows that the angular momentum of the nth orbit is just proportional to n: and Bohr knew that Planck's constant, the basis of quantum theory had the dimensions of angular momentum!

Evidently, then the angular momentum in the nth orbit was nKh, where h is Planck's constant and K is some multiplying factor, the same for all the orbits, still to be determined.

In fact, the value of K follows from the results above. R_H, m, h, and c are all known quantities (R_H being measured experimentally by observing the lines in the Balmer series) so the above formulas immediately give the electron's speed and distance from the nucleus in the nth orbit, and hence its angular momentum. Therefore, by putting in these experimentally determined quantities, we can find K.

BOHR'S SEMICLASSICAL ARGUMENT TO FIX THE QUANTUM OF ANGULAR MOMENTUM

However, Bohr found a clever *theoretical* way to determine R_H from his model: by equating his prediction of the frequency emitted when an electron goes from one orbit to another in a very large atom with the classical prediction which would be just the orbital frequency of the electron, how many times per second it goes around, he deduced $K = 1/2\pi$ and from that the Rydberg constant that appeared before is here given in terms of h, m and e. *The rather abstract argument that the quantum predictions must match the known classical results for large slow systems actually fixes the Rydberg constant.*

His argument goes as follows: for the circular orbits

$$\frac{mv^2}{r} = \frac{1}{4\pi\varepsilon_0}\cdot\frac{e^2}{r^2} \text{ so } mv^2 = \frac{1}{4\pi\varepsilon_0}\cdot\frac{e^2}{r}, \quad K.E. = -\tfrac{1}{2}P.E., \quad E = -\frac{1}{4\pi\varepsilon_0}\cdot\frac{e^2}{2r}.$$

With the angular momentum quantized, for the nth orbit:

$$mv_n r_n = nKh,$$

where h is Planck's constant, n an integer, K the unknown multiplying factor (ok, fixed by experiment, but we're finding it independently).

From this quantization condition we can find the radius, and hence the energy, of the nth orbit:

$$\frac{1}{4\pi\varepsilon_0}\cdot\frac{e^2}{r_n}=mv_n^{\,2}=m\left(\frac{nKh}{mr_n}\right)^2$$

Giving

$$r_n=\frac{4\pi\varepsilon_0 n^2K^2h^2}{me^2},\quad E_n=-\frac{1}{4\pi\varepsilon_0}\cdot\frac{e^2}{2r_n}=-\left(\frac{1}{4\pi\varepsilon_0}\right)^2\cdot\frac{me^4}{2K^2h^2}\cdot\frac{1}{n^2}.$$

In the large n limit,

$$E_{n+1}-E_n\cong\left(\frac{1}{4\pi\varepsilon_0}\right)^2\cdot\frac{me^4}{2K^2h^2}\cdot\frac{2}{n^3}=h\nu,$$

so

$$\nu=\left(\frac{1}{4\pi\varepsilon_0}\right)^2\cdot\frac{me^4}{K^2h^3n^3}$$

where v is the frequency of the emitted photon on jumping down one quantum number.

In the classical limit of large n, v *must* match the orbital frequency of the electron, since Maxwell's equations will be valid.

That is,

$$\nu=\frac{v_n}{2\pi r_n}=\frac{nKh}{2\pi mr_n^2}=\left(\frac{1}{4\pi\varepsilon_0}\right)^2\cdot\frac{nKh}{2\pi m}\cdot\frac{m^2e^4}{n^4k^4h^4}=\left(\frac{1}{4\pi\varepsilon_0}\right)^2\cdot\frac{me^4}{2\pi n^3K^3h^3}.$$

Comparing the two expressions, we see that they agree if $K=1/2\pi$

Putting $K=1/2\pi$ into the energy level formula,

$$E_n=-\left(\frac{1}{4\pi\varepsilon_0}\right)^2\cdot\frac{me^4}{2K^2h^2}\cdot\frac{1}{n^2}=-\left(\frac{1}{4\pi\varepsilon_0}\right)^2\cdot\frac{2\pi^2me^4}{h^2}\cdot\frac{1}{n^2}.$$

Now the Rydberg constant is defined by

$$E_n=-hcR_H/n^2$$

so the Bohr model *predicts* that

$$R_H = \left(\frac{1}{4\pi\varepsilon_0}\right)^2 \cdot \frac{2\pi^2 me^4}{ch^3}.$$

This formula was found to be correct within the limits of experimental error in measuring the quantities on the right. But few people believed his theory. For one thing, it soon became apparent that in the spectra of some stars (actually including the sun) there were spectral lines apparently corresponding to *half* the angular momentum quantum. How could that be? Bohr's response was that these lines must be from ionized helium, not hydrogen. A neutral helium atom has two electrons, a singly-ionized helium atom has just one electron, but the nucleus has a charge twice that of the hydrogen nucleus, so the factor e^4 in the Rydberg constant is replaced by $4e^4$, which leads to the observed result and found that actually the R_H for these new lines corresponded to a factor of 4.0016. How could Bohr explain *that*? Bohr pointed out that at this level of precision, the finite mass of the nucleus must be taken into account by using a reduced mass for the electron. This gives just the right factor. This result greatly impressed Einstein, who concluded that Bohr must be on the right track.

THE BOHR ATOM IS STILL IMPORTANT

Bohr's semiclassical analysis has long been replaced by Schrödinger's wave function, there are recent experiments in atomic physics where the classical approach provides valuable insight. In particular, so-called Rydberg atoms, which are atoms with one electron in a spatially large orbit (large n, weakly bound), act a lot like classical systems. Such atoms can be ionized by microwave fields. For a considerable range of parameters, the onset of this ionization can be accounted for by ignoring quantum mechanics altogether, and interpreting ionization as the onset of chaotic motion in the *classical* driven system! (And, the standard perturbation theoretic methods of quantum mechanics don't work for this system anyway, because the perturbing microwave electric field is of the same order of magnitude as the atom's electric field at these large orbits.) We should mention that, counterintuitively, quantum mechanics *does* become important again at *very* large n (or high microwave frequency), where some tricks from condensed matter physics have been used successfully to interpret the experiments. This is a rich subject: qualitatively different phenomena occur as the ratio of microwave frequency to orbital frequency is varied.

AN ACCIDENT AT THE PHONE COMPANY MAKES EVERYTHING CRYSTAL CLEAR

There was an accident at the Bell Telephone Laboratories in April 1925. Clinton Davisson and L. H. Germer, looking for ways to improve vacuum tubes, were watching how electrons from an electron gun in a vacuum tube scattered off a flat nickel surface. Suddenly, while the experiment was running and the nickel target was very hot, a bottle of liquid air near the apparatus exploded, smashing one of the vacuum pipes, and air rushed into the apparatus. The hot nickel target oxidized immediately. The layer of oxide made their target useless for further investigations. They decided to clean off the oxide by heating the nickel in a hydrogen atmosphere then in vacuum. After doing this for a prolonged period, the nickel looked good, and they resumed the investigation.

To their amazement, the pattern of electron scattering from the newly cleaned nickel target was completely different from that before the accident. What had changed? On examining their newly cleaned crystal carefully, they found a clue. The original target was polycrystalline made up of a multitude of tiny crystals, oriented randomly. During the prolonged heating of the cleaning process, the nickel had re-crystallized into a few large crystals.

To quote from their paper: "It seemed probable to us from these results that the intensity of scattering from a single crystal would exhibit a marked dependence on crystal direction, and we set about at once preparing experiments for an investigation of this dependence. We must admit that the results obtained in these experiments have proved to be quite at variance with our expectations. It seemed likely that strong beams would be found issuing from the crystal along what may be termed its transparent directions the directions in which the atoms in the lattice are arranged along the smallest number of lines per unit area. Strong beams are indeed found issuing from the crystal, but only when the speed of bombardment lies near one or another of a series of critical values, and then in directions quite unrelated to crystal transparency.

"The most striking characteristic of these beams is a one to one correspondence ...which the strongest of them bear to the Laue beams that would be found issuing from the same crystal if the incident beam were a beam of *x*-rays. Certain others appear to be analogues ... of optical diffraction beams from plane reflection gratings the lines of these gratings being lines or rows of atoms in the surface of the crystal. Because of these similarities ... a description ... in terms of an equivalent wave radiation ... is not only possible, but most simple and natural. This involves the association of a wavelength with the incident electron beam, and this wavelength turns out to be in acceptable agreement with the value h/mv of the undulatory mechanics, Planck's action constant divided by the momentum of the electron.

"That evidence for the wave nature of particle mechanics would be found in the reaction between a beam of electrons and a single crystal was predicted by Elsasser two years ago shortly after the appearance of L. de Broglie's original papers on wave mechanics."

It should be added that the two-slit diffraction pattern is of course exhibited by a beam of electrons, has been observed experimentally many times, and has precisely the same form as that for light. Electrons and photons generate interference patterns that are identical although the short wavelength of the electrons used presents a challenge.

Chapter 3

Wave Equations, Wavepackets and Superposition

De Broglie's doctoral thesis, defended at the end of 1924, created a lot of excitement in European physics circles. Shortly after it was published in the fall of 1925 Pieter Debye, a theorist in Zurich, suggested to Erwin Schrodinger that he give a seminar on de Broglie's work. Schrodinger gave a polished presentation, but at the end Debye remarked that he considered the whole theory rather childish: why should a wave confine itself to a circle in space? It wasn't as if the circle was a waving circular string, real waves in space diffracted and diffused, in fact they obeyed three-dimensional wave equations, and that was what was needed. This was a direct challenge to Schrodinger, who spent some weeks in the Swiss mountains working on the problem, and constructing his equation.

There is no rigorous derivation of Schrodinger's equation from previously established theory, but it can be made very plausible by thinking about the connection between light waves and photons, and construction an analogous structure for de Broglie's waves and electrons.

MAXWELL'S WAVE EQUATION

Let us examine what Maxwell's equations tell us about the motion of the simplest type of electromagnetic wave a monochromatic wave in empty space, with no currents or charges present.

As we discussed in the last lecture, Maxwell found the wave equation.

$$\nabla^2\vec{E}-\frac{1}{c^2}\frac{\partial^2\vec{E}}{\partial t^2}=0$$

which reduces to

$$\frac{\partial^2\vec{E}}{\partial x^2}-\frac{1}{c^2}\frac{\partial^2\vec{E}}{\partial t^2}=0$$

for a plane wave moving in the x-direction, with solution

$$\vec{E}(x,t)=\vec{E}_0e^{i(kx-\omega t)}$$

Applying the wave equation differential operator to this plane wave solution

$$\left(\frac{\partial^2}{dx^2}-\frac{1}{c^2}\frac{\partial^2}{\partial t^2}\right)\vec{E}_0e^{i(kx-\omega t)}=\left(k^2-\frac{\omega^2}{c^2}\right)\vec{E}_0e^{i(kx-\omega t)}=0$$

so

$\omega=ck$.

This is just the familiar statement that the wave must travel at c.

WHAT DOES THE WAVE EQUATION TELL US ABOUT THE PHOTON?

We know from the photoelectric effect and Compton scattering that the photon energy and momentum are related to the frequency and wavelength of the light by

$$E=h\nu=\hbar\omega$$

$$p=\frac{h}{\lambda}=\hbar k$$

Notice, then, that the wave equation tells us that $\omega=ck$ and hence $E = cp$.

To put it another way, if we think of $e^{i(kx-wt)}$ as describing a particle (photon) it would be more natural to write the plane wave as

$$\vec{E}_0 e^{\frac{i}{k}(px-Et)}$$

that is, in terms of the energy and momentum of the particle.

In these terms, applying the (Maxwell) wave equation operator to the plane wave yields

$$\left(\frac{\partial^2}{\partial x^2}-\frac{1}{c^2}\frac{\partial^2}{\partial t^2}\right)\vec{E}_0 e^{\frac{i}{\hbar}(px-Et)}=\left(p^2-\frac{E^2}{c^2}\right)\vec{E}_0 e^{\frac{i}{\hbar}(px-Et)}=0$$

or

$E^2 = c^2p^2$.

The wave equation operator applied to the plane wave describing the particle propagation yields the energy-momentum relationship for the particle.

CONSTRUCTING A WAVE EQUATION FOR A PARTICLE WITH MASS

The discussion above suggests how we might extend the wave equation operator from the photon case (zero rest mass) to a particle having rest mass m_0. We need a wave equation operator that, when it operates on a plane wave, yields

$$E^2 = c^2p^2 + m_0^2c^4$$

Writing the plane wave function

$$\phi(x,t) = Ae^{\frac{i}{\hbar}(pk-Et)}$$

where A is a constant, we find we can get $E^2 = c^2p^2 + m_0^2c^4$ by adding a constant (mass) term to the differentiation terms in the wave operator:

$$\left(\frac{\partial^2}{\partial x^2}-\frac{1}{c^2}\frac{\partial^2}{\partial t^2}-\frac{m_0^2c^2}{\hbar^2}\right)Ae^{\frac{i}{\hbar}(px-Et)}=-\frac{1}{\hbar^2}\left(p^2-\frac{E^2}{c^2}+m_0^2c^2\right)Ae^{\frac{i}{\hbar}(px-Et)}=0$$

This wave equation is called the *Klein-Gordon* equation and correctly describes the propagation of relativistic particles of mass m_0. However, it's a bit inconvenient for nonrelativistic particles, like the electron in the hydrogen atom, just as $E^2 = m_0^2c^4 + c^2p^2$ is less useful than $E = p^2/2m$ for this case.

A NONRELATIVISTIC WAVE EQUATION

Continuing along the same lines, let us assume that a nonrelativistic electron in free space (no potentials, so no forces) is described by a plane wave:

$$\psi(x,t) = Ae^{\frac{i}{\hbar}(px-Et)}$$

We need to construct a wave equation operator which, applied to this wave function, just gives us the ordinary nonrelativistic energy-momentum relationship, $E = p^2/2m$. The p^2 obviously comes as usual from differentiating twice with respect to x, but the only way we can get E is by having a *single* differentiation with respect to time, so this looks different from previous wave equations:

$$i\hbar\frac{\partial\psi(x,t)}{\partial t} = -\frac{\hbar^2}{2m}\frac{\partial^2\psi(x,t)}{\partial x^2}.$$

This is *Schrodinger's equation* for a free particle. It is easy to check that if $\psi(x,t)$ has the plane wave form given above, the condition for it to be a solution of this wave equation is just $E = p^2/2m$.

Notice one remarkable feature of the above equation the i on the left means that ψ *cannot* be a real function.

HOW DOES A VARYING POTENTIAL AFFECT A DE BROGLIE WAVE?

The effect of a potential on a de Broglie wave was considered by Sommerfeld in an attempt to generalize the rather restrictive conditions in Bohr's model of the atom. Since the electron was orbiting in an inverse square force, just like the planets around the sun, Sommerfeld couldn't understand why Bohr's atom had only circular orbits, no Kepler-like ellipses. (Recall that all the observed spectral lines of hydrogen were accounted for by energy differences between *circular* orbits.)

De Broglie's analysis of the allowed circular orbits can be formulated by assuming at some instant in time the spatial variation of the wave function on going around the orbit includes a phase term of the form $e^{ipq/\hbar}$, where here the parameter q measures distance around the orbit. Now for an acceptable wave function, the total phase change on going around the orbit must be $2n\pi$, where n is an integer. For the usual Bohr circular orbit, p is constant on going around, q changes by $2\pi r$, where r is the radius of the orbit, giving

$$\frac{1}{\hbar} p 2\pi r = 2n\pi, \text{so } pr = n\hbar,$$

the usual angular momentum quantization. What Sommerfeld did was to consider a general Kepler ellipse orbit, and visualize the wave going around such an orbit. Assuming the usual relationship $p = h/\lambda$ the wavelength will vary as the particle moves around the orbit, being shortest where the particle moves fastest, at its closest approach to the nucleus. Nevertheless, the phase change on moving a short distance Δq should still be $p\Delta q/\hbar$, and requiring the wave function to link up smoothly on going once around the orbit gives

$$\oint p dq = nh .$$

Thus only certain elliptical orbits are allowed. The mathematics is nontrivial, but it turns out that *every allowed elliptical orbit has the same energy as one of the allowed circular orbits.* That is why Bohr's theory gave all the energy levels. Actually, this whole analysis is old fashioned (it's called the "old quantum theory") but we've gone over it to introduce the idea of *a wave with variable wavelength, changing with the momentum as the particle moves through a varying potential.*

The reader may well be wondering at this point why it is at all useful to visualize a real wave going round an orbit, when we have stated that any solution of Schrodinger's equation is necessarily a complex function. Including those corresponding to Bohr's energy levels, in which the complex nature of the wave function only appears in a time varying phase factor, $e^{-iEt/\hbar}$. We should also add that if the spatial dependence is a real function, such as sin*kx*, it represents a standing wave, not a particle circling in one direction, which would be e^{ikx}, or $e^{-iEx/\hbar}$ Bearing all this in mind, it is still often instructive to sketch real wave functions, especially for one-dimensional problems.

SCHRODINGER'S EQUATION FOR A PARTICLE IN A POTENTIAL

Let us consider first the one-dimensional situation of a particle going in the *x*-direction subject to a "roller coaster" potential. What do we expect the wave function to look like? We would expect the wavelength to be shortest where the potential is lowest, in the valleys, because that's where the particle is going fastest maximum momentum.

With a nonzero potential present, the energy-momentum relationship for the particle becomes the energy equation

$$E = \frac{p^2}{2m} + V(x)$$

We need to construct a wave equation which leads naturally to this relationship. In contrast to the free particle cases discussed above, the relevant wave function here will no longer be a plane wave, since the wavelength varies with the potential. However, at a given *x*, the momentum is determined by the "local wavelength", that is,

$$p = -i\hbar\frac{\partial\psi}{\partial x}$$

It follows that the appropriate wave equation is:

$$i\hbar\frac{\partial\psi(x,t)}{\partial t} = -\frac{\hbar^2}{2m}\frac{\partial^2\psi(x,t)}{\partial x^2} + V(x)\psi(x,t).$$

This is the standard one-dimensional Schrodinger equation.

In three dimensions, the argument is precisely analogous. The only difference is that the square of the momentum is now a sum of three squared components, for the x, y and z directions, so ,

$$\frac{\partial^2}{\partial x^2} \text{ becomes } \frac{\partial^2}{\partial x^2}+\frac{\partial^2}{\partial y^2}+\frac{\partial^2}{\partial z^2}=\nabla^2$$

so now

$$i\hbar\frac{\partial\psi(x,y,z,t)}{\partial t}=-\frac{\hbar^2}{2m}\nabla^2\psi(x,y,z,t)+V(x,y,z)\psi(x,y,z,t).$$

This is the complete Schrodinger equation. So far, of course, it is based on plausibility arguments and hand-waving. Why should anyone believe that it really describes an electron wave? Schrodinger's test of his equation was the hydrogen atom. He looked for Bohr's "stationary states": states in which the electron was localized somewhere near the proton, and having a definite energy. The time dependence would be the same as for a plane wave of definite energy, $e^{-iEx/\hbar}$ the spatial dependence would be a time-independent function decreasing rapidly at large distances from the proton. That is, he took

$$\psi(x,y,z,t)=e^{-ZEt/\hbar}\psi(x,y,z)$$

He took advantage of the spherical symmetry by re-expressing the spatial wave function in spherical polar coordinates, and found his equation became a standard differential equation solved in the nineteenth century. The solution gave the shape of possible wave functions, and also allowed values of energy and angular momentum. These values were exactly the same as Bohr's (except that the lowest allowed state in the new theory had zero angular momentum): impressive evidence that the new theory was correct.

CURRENT CONSERVATION

When Schrodinger published this result in 1926, he also wrote down the complex conjugate equation, and proved that taking them together it was not difficult to deduce a continuity equation:

$$\frac{\partial \rho}{\partial t} + \text{div}\,\vec{j} = 0$$

$$\text{where } \rho = \psi^* \psi = |\psi|^2$$

$$\text{and } \vec{j} = \frac{\hbar}{2mi}(\psi^* \vec{\nabla} \psi - \psi \vec{\nabla} \psi^*).$$

Schrodinger believed the above continuity equations represented the conservation of electric charge, and had no further significance. He thought that after all his own equation showed the electron to be just a smooth classical wave at the deepest level. In fact, he succeeded in solving the three-dimensional equation with a Coulomb potential and he found the Bohr energy levels of the hydrogen atom. Obviously, he was on the right track! This classical revival approach, however, couldn't deal with the unpredictability of quantum mechanics, such as where a single photon or electron would land in a two-slit diffraction pattern. The truth is, Schrodinger didn't understand his own equation. Another physicist, Max Born, published a paper a few days after Schrodinger's in which he suggested that $|\psi(x,y,z,t)|^2\, dxdydz$ was the *relative probability* of finding the electron in a small volume *dxdydz* at (*x,y,z*) at time *t*. This interpretation was based directly on the analogy with light waves and photons, and has turned out to be correct. *Notation note*: ψ is called the *"amplitude"* or sometimes the *"probability amplitude"*.

PHOTONS AND ELECTRONS

We have seen that electrons and photons behave in a very similar fashion both exhibit diffraction effects, as in the double slit experiment, both have particle like or quantum behaviour. As we have already discussed, we now have a framework for understanding photons we first figure out how the electromagnetic wave propagates, using Maxwell's equations, that is, we find *E* as a function of *x,y,z,t*. Having evaluated *E*(*x,y,z,t*), the probability of finding a photon in a given small volume of space *dxdydz*, at time *t*, is proportional to $|E(x,y,z,t)|^2 dxdydz$, the energy density.

Born assumed that Schrodinger's wave function for the electron corresponded to the electromagnetic wave for the photon in the sense that the square of the modulus of the Schrodinger wave amplitude at a point was the relative probability density for finding the electron at that point. So the routine is the same: for given boundary conditions and a given potential, Schrodinger's differential equation can be solved and the wave function $\psi(x,y,z,t)$ evaluated. Then, $|\psi(x,y,z,t)|^2$ *dxdydz* gives the relative probability of finding the electron at (*x*,*y*,*z*) at time *t*.

Notice, though, that this interpretation of the wave function is *not* essential in finding the allowed energy levels in a given potential, such as the Bohr orbit energies, which Schrodinger derived before the physical significance of his wave function was understood.

How Wave Amplitude Varies in a Roller Coaster Potential

We mentioned above that for an electron traveling along a roller coaster potential, the local wavelength is related to the momentum of the electron as it passes that point. Perhaps slightly less obvious is that the *amplitude* of the wave varies: it will be largest at the tops of the hills (provided the particle has enough energy to get there) because that's where the particle is moving slowest, and therefore is most likely to be found.

Suppose following de Broglie we write down the relation between the "particle properties" of the electron and its "wave properties":

$$\frac{1}{2}mv^2 = E = hf, mv = p = h/\lambda$$

It would seem that we can immediately figure out the speed of the wave, just using $\lambda f = c'$ say. We find:

$$\lambda f = \frac{h}{mv} \cdot \frac{\frac{1}{2}mv^2}{h} = \frac{1}{2}v.$$

So the speed of the wave seems to be only half the speed of the electron! How could they stay together? What's wrong with this calculation?

LOCALIZING AN ELECTRON

To answer this question, it is necessary to think a little more carefully about the wave function corresponding to an electron traveling through a cathode ray tube, say. The electron leaves the cathode, shoots through the vacuum, and impinges on the screen. At an intermediate point in this process, it is moving through the vacuum and the wave function must be nonzero over some volume, but zero in the places the electron has not possibly reached yet, and zero in the places it has definitely left.

However, if the electron has a precise energy, say exactly a thousand electron volts, it also has a precise momentum. This necessarily implies that the wave has a precise wavelength. But the only wave with a *precise* wavelength λ has the form

$$\psi(x,t) = Ae^{i(kx-\omega t)}$$

where $k = 2\pi/\lambda$, and $\omega = 2\pi f$ The problem is that this plane sine wave extends to infinity in both spatial directions, so cannot represent a particle whose wave function is nonzero in a limited region of space. Therefore, to represent a localized particle, we must superpose waves having different wavelengths. Now, the waves representing electrons, *unlike the light waves representing photons,* travel at different speeds for different energies. Any intuition gained by thinking about superposing light waves of different wavelengths can be *misleading* if applied to electron waves. Fortunately, there are many examples in nature of waves whose speed depends on wavelength. A simple example is water waves on the ocean. We consider waves having a wavelength much shorter than the depth of the ocean. What is the ω, k relationship for these waves? We know it's not $\omega = Ck$, with a constant C, because waves of different wavelengths move at different speeds. In fact, it's easy to figure out the ω, k relationship, known as the *dispersion relation*, for these waves from a simple dimensional argument. What physical parameters can the wave frequency depend on? Obviously, the wavelength λ. We will use $k = 2\pi/\lambda$ as our variable. k has dimensions L^{-1}.

These waves are driven by gravity, so g, with dimensions LT^{-2}, is relevant. Actually, that's all. For ocean waves, surface tension is certainly negligible, as is the air density, and the water's viscosity. You might think the density of the water matters, but these waves are rather like a pendulum, in that they are driven by gravity, so increasing the density would increase both force and inertial mass by the same amount.

For these deepwater waves, then, dimensional analysis immediately gives:

$$\omega^2 = Cgk$$

where C is some dimensionless constant we cannot fix by dimensional argument, but actually it turns out to be 1.

WAVEPACKETS AND THE PRINCIPLE OF SUPERPOSITION

To return momentarily to the electron traveling through a vacuum, it is clear physically that it must have a wave function that goes to zero far away in either direction (we'll still work in one dimension, for simplicity). A localized wave function of this type is called a "wavepacket". We shall discover that a wavepacket can be constructed by adding plane waves together. Now, the plane waves we add together will individually be solutions of the Schrodinger equation.

But does it follow that the *sum* of such solutions of the Schrodinger equation is itself a solution to the equation? The answer is yes in other words, the Schrodinger equation

$$i\hbar\frac{\partial\psi(x,y,z,t)}{\partial t} = -\frac{\hbar^2}{2m}\nabla^2\psi(x,y,z,t) + V(x,y,z)\psi(x,y,z,t)$$

is a *linear* equation, that is to say, if $\psi_1(x,y,z,t)$, $\psi_2(x,y,z,t)$ are both solutions of the equation, then so is

$$\psi(x,y,z,t) = c_1\psi_1(x,y,z,t)+c_2\psi_2(x,y,z,t)$$

where c_1 and c_2 are arbitrary constants, as is easy to check. This is called the *Principle of Superposition.*

The essential point is that in Schrodinger's equation every term contains a factor ψ, but no term contains a factor ψ^2 (or a higher power). That's what is meant by a "linear" equation. If the equation *did* contain a constant term, or a term including ψ^2, superposition wouldn't work the sum of two solutions to the equation would not itself be a solution to the equation.

In fact, we have been assuming this linearity all along: when we analyse interference and diffraction of waves, we just add the two wave amplitudes at each spot. For the double slit, we take it that if the wave radiating from one slit satisfies the wave equation, then adding the two waves together will give a new wave which also satisfies the equation.

THE FIRST STEP IN BUILDING A WAVEPACKET: ADDING TWO SINE WAVES

If we add together two sine waves with frequencies close together, we get beats. This pattern can be viewed as a string of wavepackets, and is useful for gaining an understanding of why the electron speed calculated from $\lambda f = c$ above is apparently half what it should be.

We use the trigonometric addition formula:

$$\sin((k-\Delta k)x-(\omega-\Delta\omega)t)+\sin((k+\Delta k)x-(\omega+\Delta\omega)t) =$$
$$2\sin(kx-\omega t)\cos((\Delta k)x-(\Delta\omega)t).$$

This formula represents the phenomenon of beats between waves close in frequency. The first term, $\sin(kx-\omega t)$, oscillates at the average of the two frequencies. It is modulated by the slowly varying second term, often called the "envelope function", which oscillates once over a spatial extent of order $\pi/\Delta k$. This is the distance over which waves initially in phase at the origin become completely out of phase. Of course, going a further distance of order $\pi/\Delta k$ the waves will become synchronized again.

That is, beating two close frequencies together breaks up the continuous wave into a series of packets, the beats. To describe a single electron moving through space, we need a single packet. This can be achieved by superposing waves having a continuous distribution of wavelengths, or wave numbers within of order Δk, say, of k. In this case, the waves will be out of phase after a distance of order $\pi/\Delta k$ but since they have many different wavelengths, they will never get back in phase again.

PHASE VELOCITY AND GROUP VELOCITY

It will immediately become apparent that there are *two different velocities* in the dynamics: first, the velocity with which the *individual peaks* move to the right, and second the velocity at which the slowly varying envelope function the beat pattern moves. The $\lambda f = c'$individual peak velocity is determined by the term $\sin(kx - \omega t)$, it is ω/k this is called the *phase velocity*. The speed with which the beat pattern moves, on the other hand, is determined by the term $\cos((\Delta k)x-(\Delta\omega)t)$, this speed is $\Delta\omega/\Delta k = d\omega/dk$ for close frequencies. Going back one more time to the electron wavepacket, armed with this new insight, we can see immediately that the wave speed we calculated from $\lambda f = c'$ was the *phase* velocity of the waves. The packet itself will of course move at the *group* velocity and it is easy to check that this is just v.

Adding More Waves

We've seen how two sine waves of equal amplitude close together in frequency produce beats: if the waves are in phase at the origin, as we go along the x-axis they gradually fall out of phase, and cancel each other at a distance $x = \pi/2\Delta$, where 2Δ is the difference in k of the two sinkx waves. (For the moment, we are ignoring the time development of these waves: we're just looking at $t = 0$.). If we continue along the x-axis to π/Δ Now, if instead of adding *two* waves, we add *many* waves, all of different k, *but* with the k's taken from some small interval of size of order Δk and all these waves are in phase at the origin, then, again, they will all stay more or less in phase for a distance of order $x = \pi / 2\Delta$. However, as we proceed past that point, the chances of them all getting back in phase again get rapidly smaller as we increase the number of different waves. This suggests a way to construct a wavepacket: add together a lot of waves from within a narrow frequency range, and they will *only* be in phase in a region containing the origin. Adding waves in this way leads to a more general derivation of the formula $d\omega / dk$ for the group velocity. The standard approach is to replace the sum over plane waves by an integral, with the wavenumber k as the variable of integration, and the convention is to put a factor 2π in the denominator:

$$\psi(x,t) = \int_{-\infty}^{+\infty} \frac{dk}{2\pi} e^{ikx - i\omega(k)t} \phi(k)$$

Since we are constructing a wavepacket with a fairly well-defined momentum, we will take the function $\phi(k)$ to be strongly peaked at k_0, and going rapidly to zero away from that value, so the only significant contribution to the integral is from the neighbourhood of k_0. Therefore, if $\omega(k)$ is reasonably smooth (which it is) it is safe to put

$$\omega(k) = \omega(k_0) + (k - k_0)\omega'(k_0)$$

in the exponential. This gives

$$\psi(x,t) = \int_{-\infty}^{+\infty} \frac{dk}{2\pi} e^{ikx - i\omega(k_0)t - i(k-k_0)\omega'(k_0)t} \phi(k)$$

$$= e^{i(k_0 x - \omega(k_0)t)} \int_{-\infty}^{\infty} \frac{dk}{2\pi} e^{i(k-k_0)(x - \omega'(k_0)t)} \phi(k)$$

The first term just represents a single wave at k_0, and the peaks move at the phase velocity $V_{phase} = \omega / k$. The second term, the integral, is the *envelope function*: here x only appears in the combination $x - \omega'(k_0)t$ so the envelope, and hence the wavepacket, moves to the right at the group velocity:

$$v_{group} = \omega'(k_0)$$

THE GAUSSIAN WAVEPACKET

Fortunately, there is a simple explicit mathematical realization of the addition of plane waves to form a localized function: the *Gaussian wavepacket,*

$$\psi(x, t=0) = Ae^{ik_0 x} e^{-x^2/2\Delta^2}$$

where $p_0 = \hbar k_0$. For this wavepacket to represent one electron, with the probability of finding the electron in a small section of length dx at x equal to $|\psi|^2\, dx$ and the *total* probability of finding the electron somewhere equal to one, the constant A is uniquely determined (apart from a possible phase multiplier $e^{i\delta}$, which would not affect the probability). Using the standard result

$$\int_{-\infty}^{+\infty} e^{-ax^2} dx = \sqrt{\frac{\pi}{a}}$$

we find $|A|^2 = \left(\pi\Delta^2\right)^{-1/2}$ so

$$\psi(x, t=0) = \frac{1}{\left(\pi\Delta^2\right)^{1/4}} e^{ik_0 x} e^{-x^2/2\Delta^2}$$

But how do we construct this particular wavepacket by superposing plane waves? That is to say, we need a representation of the form:

$$\psi(x) = \int_{-\infty}^{+\infty} \frac{dk}{2\pi} e^{ikx} \phi(k)$$

The function $\phi(k)$ represents the weighting of plane waves in the neighbourhood of wavenumber k. This is a particular example of a *Fourier transform*. Note that if $\phi(k)$ is a bounded function, any particular k value gives a vanishingly small contribution, the plane-wave contribution to $\psi(x)$ from a range dk is $\phi(k)dk/2\pi$ In fact, $\phi(k)$ is given in terms of $\psi(x)$ by

$$\phi(k) = \int_{-\infty}^{+\infty} dx e^{-ikx} \psi(x).$$

It is perhaps worth mentioning at this point that this can be understood *qualitatively* by observing that the plane wave prefactor e^{-ikx} will interfere destructively with all plane wave components of $\psi(x)$ except that of wavenumber k, where it may at first appear that the contribution is infinite, but recall that as stated above, any particular k component has a vanishingly small weight.

In the present case, the above handwaving argument is unnecessary, because both the integrals can be carried out exactly, using the standard result:

$$\int_{-\infty}^{\infty} e^{-ax^2+bx} dx = e^{b^2/4a} \sqrt{\frac{\pi}{a}}$$

giving $\phi(k) = \left(4\pi\Delta^2\right)^{1/4} e^{-\Delta^2(k-k_0)^2/2}$. Putting this back in the integral for $\psi(x)$ shows that the integral equations are consistent. Note the normalization integrals in x-space and k-space are:

$$\int_{-\infty}^{\infty} |\psi(x)|^2 dx = 1, \quad \int_{-\infty}^{\infty} |\phi(k)|^2 \frac{dk}{2\pi} = 1.$$

The physical significance of the second equation above is that if the wavepacket goes through a diffraction grating so that the different k-components are dispersed in different directions, like the colours in white light, and a detector is arranged to register the electron if it has wavenumber between k and $k + dk$, the probability of finding it in that wavenumber range is $|\phi(k)^2|\, dk/2\pi$.

EXPECTATION VALUES AND THE UNCERTAINTY PRINCIPLE

It is clear from the expressions for $\psi(x)$ and its Fourier transform $\phi(k)$ above that the spreading of the wave function in *x*-space is inversely related to its spreading in *k*-space: the *x*-space wavefunction has spread $\sim \Delta$, the *k*-space wavefunction $\sim 1/\Delta$. This is perhaps the simplest example of Heisenberg's famous Uncertainty Principle: in quantum mechanics, both the position and momentum of a particle cannot be known precisely at the same moment; the more exactly one is specified the less well the other is known. This is an inevitable consequence of the wave nature of the probability distribution. As we have already seen, a particle with an exact momentum has a wave of specific wavelength, and the only such wave is a plane wave extending from minus infinity to infinity, so the position of the particle is completely undetermined. A particle with precisely defined position is described by a wavepacket having all wavelengths included with equal weight the momentum is completely undefined.

Definitions of $\Delta x, \Delta p$

The standard notation for the *expectation value* of an operator in a given quantum state is

$$\langle x\rangle = \int x\,|\psi(x)|^2\,dx.$$

In other words, <*x*> would be the statistical average outcome of making many measurements of *x* on identically prepared systems all in the quantum state $\psi(x)$, ignoring the time dependence here for the sake of simplicity.

When we talk about the "uncertainty" in *x*, *we mean in quantum mechanics the root mean square deviation in the measurements*. This is usually written Δx, unfortunate in view of our also standard use of Δ in the Gaussian function above, so the reader should watch carefully.

Therefore $\Delta x = \sqrt{\int (x-\langle x\rangle)^2 |\psi(x)|^2\,dx}$. For our wavepacket,

$\langle x \rangle = 0$

It is easy to check that

$$\Delta x = \frac{\Delta}{\sqrt{2}}, \quad \text{and writing } p = \hbar k, \quad \Delta p = \frac{\hbar}{\Delta\sqrt{2}} \text{ giving } \Delta x \cdot \Delta p = \frac{\hbar}{2}.$$

Chapter 4

The Uncertainty Principle

As we have shown for wavepackets, the *wave nature* of particles implies that we cannot know *both* position *and* momentum of a particle to an arbitrary degree of accuracy if Δx represents the uncertainty in our knowledge of position, and Δp that of momentum, then $\Delta p \Delta x \sim h$

where h is Planck's constant. In the real world, particles are three-dimensional and we should say $\Delta p_x \Delta x \sim h$

with corresponding equations for the other two spatial directions. The fuzziness about position is related to that of momentum *in the same direction*. Let's see how this works by trying to measure y-position and y-momentum very accurately. Suppose we have a source of electrons, say, an electron gun in a CRT (cathode ray tube, such as an old-fashioned monitor). The beam spreads out a bit, but if we interpose a sheet of metal with a slit of width w, then for particles that make it through the slit, we know y with an uncertainty $\Delta y = w$. Now, if the slit is a long way downstream from the electron gun source, we also know p_y very accurately as the electron reaches the slit, because to make it to the slit the electron's velocity would have to be aimed just right.

But does the measurement of the electron's y position in other words, having it go through the slit affect its y momentum? The answer is *yes*. If it didn't, then sending a stream of particles through the slit they would all hit very close to the same point on a screen placed further downstream. But we know from experiment that this is *not* what happens a single slit diffraction pattern builds up, of angular width $\theta \sim \lambda / w$, where the electron's de Broglie wavelength λ is given by $p_x \cong h/\lambda$ (there is a negligible contribution to λ from the y-momentum). The consequent uncertainty in p_y is $\Delta py / px \sim \theta \sim \lambda / w$ Putting in $p_x = h/\lambda$, we find immediately that $\Delta p_y \sim h/w$ so the act of measuring the electron's y position has fuzzed out its y momentum by precisely the amount required by the uncertainty principle.

TRYING TO BEAT THE UNCERTAINTY PRINCIPLE

In order to understand the Uncertainty Principle better, let's try to see what goes wrong when we actually try to measure position and momentum more accurately than allowed. For example, suppose we look at an electron through a microscope. What could we expect to see? Of course, you know that if we try to look at something *really* small through a microscope it gets blurry a small sharp object gets diffraction patterns around its edges, indicating that we are looking at something of size comparable to the wavelength of the light being used. If we look at something much smaller than the wavelength of light like the electron we would expect a diffraction pattern of concentric rings with a circular blob in the middle. The size of the pattern is of order the wavelength of the light, in fact from optics it can be shown to be $\sim \lambda f / d$ where d is the diameter of the object lens of the microscope, f the focal distance (the distance from the lens to the object). We shall take $f/d \sim 1$, as it usually is. So looking at an object the size of an electron should give a diffraction pattern centered on the location of the object. That would seem to pin down its position fairly precisely.

What about the *momentum* of the electron? Here a problem arises that doesn't matter for larger objects the light we see has, of course, bounced off the electron, and so the electron has some recoil momentum. That is, by bouncing light off the electron we have given it some momentum. Can we say how much? To make it simple, suppose we have good eyes and only need to bounce one photon off the electron to see it. We know the initial momentum of the photon (because we know the direction of the light beam we're using to illuminate the electron) and we know that after bouncing off, the photon hits the object lens and goes through the microscope, but we don't know *where* the photon hit the object lens. The whole point of a microscope is that all the light from a point, light that hits the object lens *in different places*, is all focused back to one spot, forming the image (apart from the blurriness mentioned above). So if the light has wavelength λ, its constituent photons have momentum $\sim h/\lambda$, and from our ignorance of where the photon entered the microscope we are uncertain of its x-direction momentum by an amount $\sim h/\lambda$. Necessarily, then, we have the same uncertainty about the electron's x-direction momentum, since this was imparted by the photon bouncing off.

But now we have a problem. In our attempts to minimize the uncertainty in the electron's momentum, by only using one photon to detect it, we are not going to see much of the diffraction pattern discussed above such diffraction patterns are generated by *many* photons hitting the film, retina or whatever detecting equipment is being used. A single photon generates a single point (at best!). This point will most likely be within of order λ of the center of the pattern, but this leaves us with an uncertainty in position of order λ.

Therefore, in attempting to observe the position and momentum of a single electron using a single photon, we find an uncertainty in position $\Delta x \sim \lambda$, and in momentum $\Delta p_x \sim h/\lambda$. These results are in accordance with Heisenberg's Uncertainty Principle $\Delta x . \Delta p_x \sim h$.

Of course, we could pin down the position much better if we used N photons instead of a single one. From statistical theory, it is known that the remaining uncertainty $\sim \lambda/\sqrt{N}$. But then N photons have bounced off the electron, so, since each is equally likely to have gone through any part of the object lens, the uncertainly in momentum of the electron as a consequence of these collisions goes up as $\sqrt{N}$. (The same as the average imbalance between heads and tails in a sequence of N coin flips.)

Noting that the uncertainty in the momentum of the electron arises because we don't know where the bounced-off photon passes through the object lens, it is tempting to think we could just use a smaller object lens, that would reduce Δp_x. Although this is correct, recall from above that we stated the size of the diffraction pattern was $\sim \lambda f/d$, where d is the diameter of the object lens and f its focal length. It is easy to see that the diffraction pattern, and consequently Δx, gets bigger by just the amount that Δp_x gets smaller!

ELECTRONS IN THE DOUBLE SLIT EXPERIMENT

Suppose now that in the double slit experiment, we set out to detect which slit each electron goes through by shining a light just behind the screen and watching for reflected light from the electron immediately after it had passed through a slit. Following the discussion in Feynman's *Lectures in Physics*, Volume III, we shall now establish that if we can detect the electrons, we ruin the diffraction pattern! Taking the distance between the two slits to be d, the dark lines in the diffraction pattern are at angles $\left(n+\frac{1}{2}\right)\lambda_{elec} = d\sin\theta$. If the light used to see which slit the electron goes through generates an uncertainty in the electron's y momentum Δp_y, in order not to destroy the diffraction pattern we must have $\Delta p_y / p < \lambda_{elec}/d$ (the angular uncertainty in the electron's direction must not be enough to spread it from the diffraction pattern maxima into the minima). Here p is the electron's full momentum, $p = h/\lambda_{elec}$. Now, the uncertainty in the electron's y momentum, looking for it with a microscope, is $\Delta p_y \sim h/\lambda_{light}$. Substituting these values in the inequality above we find the condition for the diffraction pattern to survive is $\lambda_{light} > d$, the wavelength of the light used to detect which slit the electron went through must be greater than the distance between the slits. Unfortunately, the light scattered from the electron then gives one point in a diffraction pattern of size the wavelength of the light used, so *even if we see the flash this does not pin down the electron sufficiently to say which slit it went through.* Heisenberg wins again.

UNCERTAINTY PRINCIPLE DETERMINES THE SIZE OF EVERYTHING

It is interesting to see how the actual physical size of the hydrogen atom is determined by the wave nature of the electron, in effect, by the Uncertainty Principle. In the ground state of the hydrogen atom, the electron minimizes its total energy. For a classical atom, the energy would be minus infinity, assuming the nucleus is a point (and very large in any case) because the electron would sit right on top of the nucleus. However, this cannot happen in quantum mechanics. Such a very localized electron would have a very large uncertainty in momentum in other words, the kinetic energy would be large. This is most clearly seen by imagining that the electron is going in a circular orbit of radius *r* with angular momentum $h/2\pi$. Then one wavelength of the electron's de Broglie wave just fits around the circle, $\lambda_{elec} = 2\pi r$. Clearly, as we shrink the circle's radius *r*, λ_{elec} goes down proportionately, and the electrons momentum $p = h/\lambda_{elec} = h/2\pi r$ increases. Adding the electron's electrostatic potential energy we find the total energy for a circular orbit of radius *r* is:

$$E(r) = K.E. + P.E. = \frac{p^2}{2m} - \frac{e^2}{4\pi\varepsilon_0 r} = \frac{h^2}{8m\pi^2 r^2} - \frac{e^2}{4\pi\varepsilon_0 r}$$

Notice that for very large *r*, the potential energy dominates, the kinetic energy is negligible, and shrinking the atom lowers the total energy. However, for small enough *r*, the (always positive) kinetic energy term wins, and the total energy *grows* as the atom shrinks. Evidently, then, there must be a value of *r* for which *the total energy is a minimum.* Visualizing a graph of the total energy given by the equation above as a function of *r*, at the minimum point the slope of is zero, E(r).

$dE(r)/dr = 0$

That is, $$-\frac{h^2}{4m\pi^2 r}+\frac{e^2}{4\pi\varepsilon_0}=0$$

giving. $$r_{\min}=\frac{\varepsilon_0 h^2}{\pi m e^2}$$

The total energy for this radius is the exact right answer, which is reassuring.The point of this exercise is to see that in quantum mechanics, unlike classical mechanics, a particle cannot position itself at the exact minimum of potential energy, because that would require a very narrow wave packet and thus be expensive in kinetic energy. The ground state of a quantum particle in an attractive potential is a trade off between potential energy minimization and kinetic energy minimization. Thus the physical sizes of atoms, molecules and ultimately ourselves are determined by Planck's constant.

Chapter 5

Electron in a Box

The best way to gain understanding of Schrodinger's equation is to solve it for various potentials. The simplest is a one-dimensional "particle in a box" problem. The appropriate potential is $V(x) = 0$ for x between 0, L and $V(x)$ = infinity otherwise that is to say, there are infinitely high walls at $x = 0$ and $x = L$, and the particle is trapped between them. This turns out to be quite a good approximation for electrons in a long molecule, and the three-dimensional version is a reasonable picture for electrons in metals.

Between $x = 0$ and $x = L$ we have $V = 0$, so the wave equation is just

$$i\hbar \frac{\partial \psi(x,t)}{\partial t} = -\frac{\hbar^2}{2m} \frac{\partial^2 \psi(x,t)}{\partial x^2}$$

A possible plane wave solution is

$$\psi(x,t) = Ae^{\frac{i}{\hbar}(px-Et)}$$

On inserting this into the zero-potential Schrodinger equation above we find $E = p^2/2m$, as we expect. It is very important to notice that the complex conjugate, proportional to $e^{-\frac{i}{\hbar}(px-Et)}$, is *not* a solution to the Schrodinger equation! If we blindly put it into the equation we get $E = -p^2/2m$, an unphysical result. However, a wave function proportional to gives $E = p^2/2m$, so this plane wave is a solution to the equation.

Therefore, the two allowed plane-wave solutions to the zero-potential Schrodinger equation are proportional to $e^{\frac{i}{\hbar}(px-Et)}$ and $e^{-\frac{i}{\hbar}(px+Et)}$ respectively. Note that these two solutions have the *same time dependence* $e^{\frac{-iEt}{\hbar}}$.

To decide on the appropriate solution for our problem of an electron in a box, of course we have to bring in the walls what they mean is that $\kappa = 0$ for $x < 0$ and for $x > L$ because remember $|\psi|^2$ tells us the probability of finding the particle anywhere, and, since it's in the box, it's trapped *between* the walls, so there's zero probability of finding it outside.The condition $\psi = 0$ at $x = 0$ and $x = L$ reminds us of the vibrating string with two fixed ends the solution of the string wave equation is standing waves of sine form. In fact, taking the difference of the two permitted plane-wave forms above gives a solution of this type:

$$\psi(x,t) = A\sin\frac{px}{\hbar}e^{-\frac{iEt}{\hbar}}$$

This wave function satisfies the Schrodinger equation between the walls, it vanishes at the $x = 0$ wall, it will also vanish at $x = L$ provided that the momentum variable satisfies:

$$\frac{pL}{\hbar} = \pi,\ 2\pi,\ 3\pi...$$

Thus the allowed values of p are $hn/2L$, where $n = 1, 2, 3...$, and from $E = p^2/2m$ the allowed energy levels of the particle are:

$$E = \frac{p^2}{2m} = \frac{1}{2m}\left(\frac{h}{2L}\right)^2, \frac{4}{2m}\left(\frac{h}{2L}\right)^2, \frac{9}{2m}\left(\frac{h}{2L}\right)^2, ...$$

Note that these energy levels become more and more widely spaced out at high energies, in contrast to the hydrogen atom potential.

What about the overall multiplicative constant A in the wave function? This can be real or complex. To find its value, note that at a fixed time, say $t = 0$, the probability of the electron being between x and $x + dx$ is or

$$|\psi(x)|^2 \, dx$$

$$|A|^2 \sin^2 \frac{px}{\hbar} dx.$$

The *total* probability of the particle being *somewhere* between 0, L must be unity so

$$\int_{x=0}^{x=L} |A|^2 \sin^2 \frac{px}{\hbar} dx = 1, \text{ so } \frac{1}{2} L |A|^2 = 1$$

Hence

$$\psi(x,t) = \sqrt{\frac{2}{L}} \sin \frac{px}{\hbar} e^{-\frac{iEt}{\hbar}}$$

When A is fixed in this way, by demanding that the total probability of finding the particle somewhere be unity, it is called the *normalization constant.*

STATIONARY STATES

Notice that at a later time the probability distribution for the wave function

$$\psi(x,t) = \sqrt{\frac{2}{L}} \sin \frac{px}{\hbar} e^{-\frac{iEt}{\hbar}}$$

is *the same,* because *time only appears as a phase factor* in this time-dependent function, and so does not affect $|\psi(x)|^2$. A state with a time-independent probability distribution is called a *stationary state.* Such a state must have a defin te energy as can be seen by constructing a solution to Schrodinger's equation by adding states of different energies, for example:

$$\psi(x,t) = \sqrt{\frac{1}{L}} \left(\sin \frac{\pi x}{L} e^{-\frac{i\pi\hbar t}{4mL^2}} + \sin \frac{2\pi x}{L} e^{-\frac{i\pi\hbar t}{mL^2}} \right).$$

(You can check the normalization constant at $t = 0$). For general x, the two terms in the bracket rotate in the complex plane at different rates, so their sum has a time-varying magnitude. That is to say, $|\psi(x)|^2$ varies in time, so the particle must be moving around this is *not* a stationary state. Of course, the *total* probability of finding the particle *somewhere* in the box remains unity: the normalization constant is time-independent.

THE TIME-INDEPENDENT SCHRODINGER EQUATION: EIGENSTATES AND EIGENVALUES

In contrast to the superposition of different energy states in the equation above, a stationary state has a precisely defined energy, and the energy only appears in the overall phase factor,

$$\psi(x,t) = \psi(x)e^{-\frac{iEt}{\hbar}}$$

Putting this wave function into the Schrodinger equation we find

$$-\frac{\hbar^2}{2m}\frac{\partial^2\psi(x)}{\partial x^2} + V(x)\psi(x) = E\psi(x).$$

This is the *time-independent Schrodinger equation,* and its solutions are the spatial wave functions for stationary states, states of definite energy. These are often called *eigenstates* of the equation.

The values of energy corresponding to these eigenstates are called the *eigenvalues*.

Consider again the wavefunction for the lowest energy state of a particle confined between walls at $x = 0$ and $x = L$. The reader should sketch the wavefunction from some point to the left of $x = 0$ over to the right of $x = L$. To the left of $x = 0$, the wavefunction is exactly zero, then at $x = 0$ it takes off to the right (inside the box) as a sine curve. In other words, at the origin the slope of the wavefunction ψ is zero to the left, nonzero to the right. *There is a discontinuity in the slope at the origin: this means the second derivative of* ψ *is* infinite *at the origin.* On examining the time-independent Schrodinger equation above, we see the equation can *only* be satisfied at the origin because the potential becomes infinite there the wall is an infinite potential. (And, in fact, since ψ becomes zero on approaching the origin from inside the box, the limit must be treated carefully.)

It now becomes obvious that if the box does not have infinite walls, but merely high ones, ψ describing a confined particle cannot suddenly go to zero at the walls: the second derivative must remain finite. For non-infinite walls, ψ and its derivative must be continuous on entering the wall. This has the important physical consequence that ψ will be nonzero at least for some distance into the wall, even if classically the confined particle does not have enough energy to "climb the wall". (Which it doesn't, if it's confined.) Thus, in quantum mechanics, there is a non-vanishing probability of finding the particle in a region which is "classically forbidden" in the sense that it doesn't have enough energy to get there.

Chapter 6

Fourier Series, Fourier Transforms and the Delta Function

We begin with a brief review of Fourier series. Any periodic function of interest in physics can be expressed as a series in sines and cosines we have already seen that the quantum wave function of a particle in a box is precisely of this form. The important question in practice is, for an arbitrary wave function, how good an approximation is given if we stop summing the series after N terms. We establish here that the sum after N terms, $f_N(\theta)$, can be written as a convolution of the original function with the function $\delta_N(x) = (1/2\pi)\left(\sin\left(N + \frac{1}{2}\right)x\right) / \sin\frac{1}{2}x$,, that is, $f_N(\theta) = \int_{-x}^{x} \delta_N(\theta - \theta') f(\theta') d\theta'$. The structure of the function $\delta_N(x)$, when put together with the function $f(\theta)$, gives a good intuitive guide to how good an approximation the sum over N terms is going to be for a given function $f(\theta)$. In particular, it turns out that step discontinuities are *never* handled perfectly, no matter how many terms are included. Fortunately, *true* step discontinuities never occur in physics, but this is a warning that it is of course necessary to sum up to some N where the sines and cosines oscillate substantially more rapidly than any sudden change in the function being represented.

We go on to the Fourier transform, in which a function on the *infinite* line is expressed as an integral over a *continuum* of sines and cosines (or equivalently exponentials e^{ikx}). It turns out that arguments analogous to those that led to $\delta_N(x)$ now give a function $\delta(x)$ such that

$$f(x) = \int_{-\infty}^{\infty} \delta(x - x') \; f(x')dx'$$

Confronted with this, one might well wonder what is the point of a function $\delta(x)$ which on convolution with $f(x)$ gives back the same function $f(x)$. The relevance of $\delta(x)$ will become evident later, when states of a quantum particle are represented by wave functions on the infinite line, like $f(x)$, and operations on them involve integral operators similar to the convolution above. Working with operations on these functions is the continuum generalization of matrices acting on vectors in a finite-dimensional space, and $\delta(x)$ is the infinite-dimensional representation of the unit matrix. Just as in matrix algebra the eigenstates of the unit matrix are a set of vectors that span the space, and the unit matrix elements determine the set of dot products of these basis vectors, the delta function determines the generalized inner product of a *continuum* basis of states. It plays an essential role in the standard formalism for continuum states, and you need to be familiar with it!

FOURIER SERIES

Any *reasonably smooth* real function $f(\theta)$ defined in the interval $-\pi < \theta \leq \pi$ can be expanded in a Fourier series,

$$f(\theta) = \frac{A_0}{2} + \sum_{n=1}^{\infty} (A_n \cos n\theta + B_n \sin n\theta)$$

where the coefficients can be found using the orthogonality condition,

$$\int_{-\pi}^{\pi} \cos m\theta \cos n\theta d\theta = \pi\delta_{m,n}$$

and the same condition for the $\sin n\theta$'s to give:

$$A_n = \frac{1}{\pi}\int_{-\pi}^{\pi} f(\theta)\cos n\theta d\theta$$

$$B_n = \frac{1}{\pi}\int_{-\pi}^{\pi} f(\theta)\sin n\theta d\theta$$

Note that for an *even* function, only the A_n are nonzero, for an odd function only the B_n are nonzero.

How Smooth is "Reasonably Smooth"?

The number of terms of the series necessary to give a good approximation to a function depends on how rapidly the function changes. To get an idea of what goes wrong when a function is not "smooth", it is instructive to find the Fourier sine series for the step function

$f(\theta) = -1$ for $-\pi < \theta \le 0$, $\quad f(\theta) = 1$ for $0 < \theta \le \pi$.

Using the expression for B_n above it is easy to find:

$$f(\theta) = \frac{4}{\pi}\left(\sin\theta + \frac{\sin 3\theta}{3} + \frac{\sin 5\theta}{5} + \ldots\right)$$

Taking the first half dozen terms in the series gives:

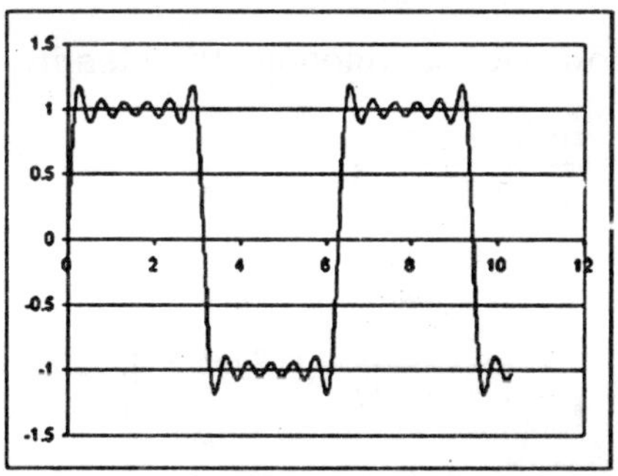

As we include more and more terms, the function becomes smoother but, surprisingly, *the initial overshoot at the step stays at a finite fraction of the step height*. However, the function recovers more and more rapidly, that is to say, the overshoot and "ringing" at the step take up less and less space. This overshoot is called *Gibbs' phenomenon*, and only occurs in functions with discontinuities.

How the Sum over N Terms is Related to the Complete Function. To get a clearer idea of how a Fourier series converges to the function it represents, it is useful to stop the series at *N* terms and examine how that sum, which we denote $f_N(\theta)$, tends towards $f(\theta)$.

So, substituting the values of the coefficients

$$A_n = \frac{1}{\pi}\int_{-\pi}^{\pi} f(\theta)\cos n\theta d\theta$$

$$B_n = \frac{1}{\pi}\int_{-\pi}^{\pi} f(\theta)\sin n\theta d\theta$$

in the series

$$f_N(\theta) = \frac{A_0}{2} + \sum_{n=1}^{N}(A_n \cos n\theta + B_n \sin n\theta)$$

gives

$$f_N(\theta) = \frac{1}{2\pi}\int_{-\pi}^{\pi} f(\theta')d\theta' + \frac{1}{\pi}\sum_{n=1}^{N}\int_{-\pi}^{\pi}(\cos n\theta\cos n\theta' + \sin n\theta \sin n\theta')f(\theta')d\theta'$$

$$= \frac{1}{2\pi}\int_{-\pi}^{\pi} f(\theta')d\theta' + \frac{1}{\pi}\sum_{n=1}^{N}\int_{-\pi}^{\pi}\cos n(\theta-\theta')f(\theta')d\theta'.$$

We can now use the trigonometric identity

$$\sum_{n=1}^{N}\cos nx = \frac{\sin(N+\frac{1}{2})x}{2\sin\frac{1}{2}x} - \frac{1}{2}$$

to find

$$f_N(\theta) = \frac{1}{2\pi}\int_{-\pi}^{\pi}\frac{\sin(N+\frac{1}{2})(\theta-\theta')}{\sin\frac{1}{2}(\theta-\theta')}f(\theta')d\theta' = \int_{-\pi}^{\pi}\delta_N(\theta-\theta')f(\theta')d\theta'$$

where

$$\delta_N(x) = \frac{1}{2\pi}\frac{\sin(N+\frac{1}{2})x}{\sin\frac{1}{2}x}.$$

(Note that proving the trigonometric identity is straightforward: write $z = e^{ix}$, so $\cos nx = \frac{1}{2}(z^n + z^{-n})$, and sum the geometric progressions.)

Going backwards for a moment and writing

$$\delta_N(x) = \frac{1}{\pi}\left(\sum_{n=1}^{N} \cos nx + \frac{1}{2}\right)$$

it is easy to check that

$$\int_{-\pi}^{\pi} \delta_N(x)dx = 1.$$

To help visualize $\delta_N(\theta)$, here is $N = 20$:

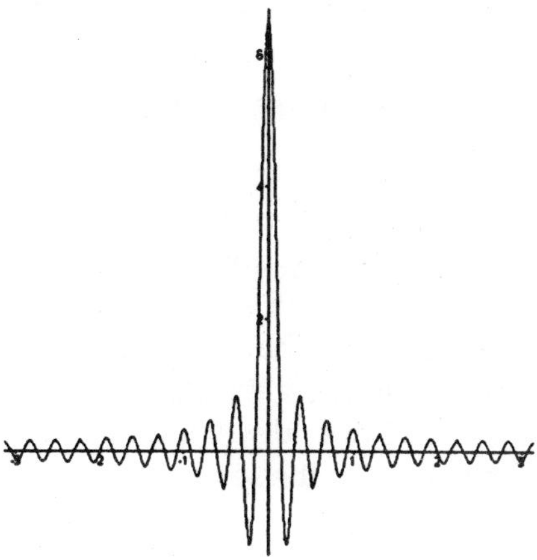

We have just established that the total area under the curve = 1, and it is clear from the diagram that almost all this area is under the central peak, since the areas away from the centre are almost equally positive and negative. The width of the central peak is $\pi/(N+\frac{1}{2})$, its height $(N+\frac{1}{2})/\pi$. For functions *varying slowly* compared with the oscillations, the convolution integral

$$f_N(\theta) = \int_{-\pi}^{\pi} \delta_N(\theta-\theta')f(\theta')d\theta'$$

will give $f_N(\theta)$ close to $f(\theta)$, and for these functions $f_N(\theta)$ will tend to $f(\theta)$ as N increases.

It is also clear why convoluting this curve with a step function gives an overshoot and oscillations. Suppose the function $f(\theta)$ is a step, jumping from 0 to 1 at $\theta = 0$ From the convolutionary form of the integral, you should be able to convince yourself that the value of $f_N(\theta)$ at a point θ is the total area under the curve to the *left* of that point (area below zero—that is, below the x-axis—of course counting negative). For $\theta = 0$, this must be exactly 0.5 (since all the area under $\delta_N(\theta)$ adds to 1). *But* if we want the value of $f_N(\theta)$ at $\theta = \pi/(2N+1)$ (that is, the first point to the right of the origin where the curve cuts through the x-axis), we must *add* all the area to the left of $\theta = \pi/(2N+1)$, which actually adds up to a total area greater than one, since the leftover area to the right of that point is overall negative. That gives the overshoot.

A FOURIER SERIES IN QUANTUM MECHANICS: ELECTRON IN A BOX

The time-independent Schrodinger wave functions for an electron in a box (here a one-dimensional square well with infinite walls) are just the sine and cosine series determined by the boundary conditions. Therefore, any reasonably smooth initial wavefunction describing the electron can be represented as a Fourier series. The time development can then be found be multiplying each term in the series by the appropriate time-dependent phase factor.

Exercise: prove that for a function $f(\theta) = \sum_{n=-\infty}^{\infty} a_n e^{in\theta}$, with the a_n in general complex,

$$\frac{1}{2\pi}\int_{-\pi}^{\pi} |f(\theta)|^2 \, d\theta = \sum_{n=-\infty}^{\infty} |a_n|^2.$$

(The physical relevance of this result is as follows: for an electron confined to the circumference of a ring of unit radius, θ is the position of the electron. An orthonormal basis of states of the electron on this ring is the set of functions $\left(1/\sqrt{2\pi}\right)e^{in\theta}$ with n an integer, a correctly normalized superposition of these states must have $\sum_{n\to\infty}^{\infty} |a_n|^2 = 1$, so that the total probability of finding the electron in some state is unity. But this must also mean that the total probability of finding the electron anywhere on the ring is unity—and that's the left-hand side of the above equation—the 2π's cancel.)

EXPONENTIAL FOURIER SERIES

In the previous lecture, we discussed briefly how a Gaussian wave packet in x-space could be represented as a continuous linear superposition of plane waves that turned out to be another Gaussian wave packet, this time in k-space. The plan here is to demonstrate how we can arrive at that representation by carefully taking the limit of the well-defined Fourier series, going from the finite interval $(-\pi,\pi)$ to the whole line, and to outline some of the mathematical problems that arise, and how to handle them.

The first step is a trivial one: we need to generalize from real functions to complex functions, to include wave functions having nonvanishing current. A smooth *complex* function can be written in a Fourier series simply by allowing A_n and B_n to be complex, but in this case a more natural expansion would be in powers of $e^{i\theta}, e^{-i\theta}$

We write:

$$f(\theta)=\sum_{n=-\infty}^{\infty} a_n e^{in\theta} \quad \text{with } a_n=\frac{1}{2\pi}\int_{-\pi}^{\pi} e^{-in\theta'} f(\theta')d\theta'$$

and retracing the above steps

$$f_N(\theta)=\frac{1}{2\pi}\int_{-\pi}^{\pi}\sum_{n=-N}^{N} e^{in(\theta-\theta')} f(\theta')d\theta'$$

$$=\frac{1}{2\pi}\int_{-\pi}^{\pi} f(\theta')d\theta'+\frac{1}{\pi}\int_{-\pi}^{\pi}\sum_{n=1}^{N}\cos n(\theta-\theta')f(\theta')d\theta'$$

exactly the same expression as before, therefore giving the same $\delta_N(\theta)$. This isn't surprising, because using $\cos n\theta=\frac{1}{2}\left(e^{in\theta}+e^{-in\theta}\right)$, $\sin n\theta=\frac{1}{2i}\left(e^{in\theta}-e^{-in\theta}\right)$ the first N terms in A_n, B_n can simply be rearranged to a sum over $e^{in\theta}$ terms for $-N\le n\le N$.

ELECTRON OUT OF THE BOX: THE FOURIER TRANSFORM

To break down a *wave packet* into its plane wave components, we need to extend the range of integration from the $(-\pi, \pi)$ used above to $(-\infty,\infty)$. We do this by first rescaling from $(-\pi, \pi)$ to $(-L/2, L/2)$ and then taking the limit $L\rightarrow\infty$.

Scaling the interval from 2π to L (in the complex representation) gives:

$$f(x)=\sum_{n=-\infty}^{\infty} a_n e^{2\pi i n x/L} \text{ where } a_n=\frac{1}{L}\int_{-L/2}^{L/2} f(x)e^{-2\pi i n x/L}dx$$

the sum in n being over all integers. This is an expression for $f(x)$ in terms of plane waves e^{ikx} where the allowed k's are $2\pi n/l$, with $n = 0, \pm1, \pm2, \ldots$

Retracing the steps above in the derivation of the function $\delta_N(x)$, we find the equivalent function to be

$$\delta_N^L(x)=\frac{1}{L}\left(1+2\sum_{n=1}^{N}\cos\frac{2\pi n x}{L}\right)=\frac{\sin\left((2N+1)\pi x/L\right)}{L\sin\left(\pi x/L\right)}.$$

Studying the expression on the right, it is evident that *provided N is much greater than L,* this has the same peaked-at-the-origin behaviour as the $\delta_N(x)$ we considered earlier. *But* we are interested in the limit $L\rightarrow\infty$, and there—for fixed N—this function $\delta_N^L(x)$ is low and flat.

Therefore, we must take the limit N going to infinity before taking L going to infinity.

This is what we do in the rest of this section. Provided L is finite, we still have a Fourier series, representing a function of period L. Our main interest in taking L infinite is that we would like to represent a *nonperiodic* function, for example a localized wave packet, in terms of plane-wave components.

Suppose we have such a wave packet, say of length L_1, by which we mean the wave is exactly zero outside a stretch of the axis of length L_1. Why not just express it in terms of an infinite N Fourier series based on some large interval $(-L/2,L/2)$ provided the wave packet length L_1 is completely inside this interval? The point is that such an analysis would indeed accurately reproduce the wave packet inside the interval, but the *same* sum of plane waves evaluated *over all the x-axis* would reveal an *infinite string* of identical wave packets L apart! This is not what we want.

As a preliminary to taking L to infinity, let us write the exponential plane wave terms in the standard k-notation,

$$e^{2\pi i n x/L} = e^{ik_n x}$$

So we are summing over an (infinite N) set of plane waves having wave number values

$$k_n = 2\pi n/L,\ n = 0, \pm 1, \pm 2, \ldots,$$

a set of equally-spaced k's with separation $\Delta k = 2\pi/L$.

Consider now what happens if we *double* the basic interval from (-L/2, L/2) to (-L, L).

The new allowed k values are $k_n = \pi n/L,\ n = 0, \pm 1, \pm 2, \ldots$, so the separation is now $\Delta k = \pi/L$, *half* of what it was before. It is evident that as we increase L, the spacing between successive k_n values gets less and less.

Going back to the interval of length L, writing $La_n = a(k_n), k_n = 2\pi n/L$ we have

$$f(x) = \sum_{n=-\infty}^{\infty} a_n e^{2\pi i n x/L} = \frac{1}{L}\sum_{n=-\infty}^{\infty} a(k_n) e^{ik_n x}.$$

Recall that the Riemann integral can be defined by

$$\int f(k)dk = \lim_{\Delta k \to 0} \sum f(k_n)\Delta k$$

with $k_n = n\Delta k,\ n = 0, \pm 1, \pm 2, \ldots.$

The expression on the right-hand side of the equation for $f(x)$ has the same form as the right-hand side of the Riemann integral definition, and here $\Delta k = 2\pi/L$. That is to say,

$$2\pi f(x) = \frac{2\pi}{L}\sum_{n=-\infty}^{\infty} a(k_n)e^{ik_n x} = \sum_{n=-\infty}^{\infty} a(k_n)e^{ik_n x}\Delta k = \int_{-\infty}^{\infty} a(k)e^{ikx}dk$$

in the limit $\Delta k \to 0$, or equivalently $L \to \infty$. We are of course assuming here that the function $a(k_n)$, which we have only defined (for a given L) on the set of points $k_{n'}$, tends to a continuous function $a(k)$ in the limit $L \to \infty$.

It follows that in the infinite L limit, we have the *Fourier transform* equations:

$$f(x) = \frac{1}{2\pi}\int_{-\infty}^{\infty} a(k)e^{ikx}dk \text{ and } a(k) = \lim_{L\to\infty} La_n$$

$$= \lim_{L\to\infty} \int_{-L/2}^{L/2} f(x)e^{-2\pi inx/L}dx = \int_{-\infty}^{\infty} f(x)e^{-ikx}dx$$

DIRAC'S DELTA FUNCTION

Now we have taken both N and L to infinity, what has happened to our function $\delta_N^L(x)$? Remember that our procedure for finding $f_N(\theta)$ in terms of $f(\theta)$ gave the equation

$$f_N(\theta) = \frac{1}{2\pi}\int_{-\pi}^{\pi} f(\theta')d\theta' + \frac{1}{\pi}\int_{-\pi}^{\pi}\sum_{n=1}^{N}\cos n(\theta-\theta')f(\theta')d\theta'$$

and from this we found $\delta_N(\theta)$.

Following the same formal procedure with the ($L = \infty$) Fourier transforms, we are forced to take N infinite (recall the procedure only made sense if N was taken to infinity before L), so in place of an equation for $f_N(\theta)$ in terms of $f(\theta)$, we get an equation for $f(x)$ in terms of itself! Let's write it down first and think afterwards:

$$f(x) = \frac{1}{2\pi}\int_{-\infty}^{\infty}\left(\int_{-\infty}^{\infty} f(x')e^{-ikx'}dx'\right)e^{ikx}dk = \int_{-\infty}^{\infty}\left(\int_{-\infty}^{\infty}\frac{dk}{2\pi}e^{ik(x-x')}\right)f(x')dx'$$

In other words,

$$f(x) = \int_{-\infty}^{\infty} \delta(x-x')f(x')dx'$$

where

$$\delta(x) = \int_{-\infty}^{+\infty}\frac{dk}{2\pi}e^{ikx}.$$

This is the *Dirac delta function*. This hand-waving approach has given a result which is not clearly defined. The function is linearly divergent at the origin, and has finite oscillatory behaviour everywhere else. To make any progress, we must provide some form of cutoff in *k*-space, then perhaps we can find a meaningful limit by placing the cutoff further and further away.

From our arguments above, we should be able to recover $\delta(x)$ as a limit of $\delta_N^L(x)$ by first taking *N* to infinity, then *L*. That is to say,

$$\delta(x) = \lim_{L\to\infty}\left(\lim_{N\to\infty}\delta_N^L(x)\right) = \lim_{L\to\infty}\left(\lim_{N\to\infty}\frac{\sin\left((2N+1)\pi x/L\right)}{L\sin(\pi x/L)}\right).$$

A way to understand this limit is to write $M = (2N+1)\pi/L$ and let *M* go to infinity before *L*. (This means as we take *L* large on its way to infinity, we're taking *N* far larger!)

So the numerator is just Mx. In the limit of infinite *L*, for any finite *x* the denominator is just πx, since $\sin\theta = \theta$ in the limit of small θ.

From this,

$$\delta(x) = \lim_{M\to\infty}\frac{\sin Mx}{\pi x}.$$

This is still a rather pathological function, in that it is oscillating more and more quickly as the infinite limit is taken. This comes about from the abrupt cutoff in the sum at the frequency *N*.

To see how this relates to the (also ill-defined) $\delta(x) = \int_{-\infty}^{+\infty}(dk/2\pi)e^{ikx}$, recall $\delta_N^L(x)$ came from the series

$$\delta_N^L(x) = \frac{\sin(2N+1)\pi x/L}{L\sin\pi x/L} = \frac{1}{L}\left(1+2\sum_{n=1}^{N}\cos\frac{2\pi nx}{L}\right).$$

Expressing the cosine in terms of exponentials, then replacing the sum by an integral in the large *N* limit, in the same way we did earlier, writing $k_n = 2\pi n/L$, so the interval between successive k_n 's is $\Delta k = 2\pi/L$, so $\int f(k)dk \cong (2\pi/L)\sum f(k_n)$:

$$\delta_N^L(x) = \frac{1}{L}\sum_{n=-N}^{N} e^{2\pi inx/L} \cong \int_{-2\pi N/L}^{2\pi N/L} \frac{dk}{2\pi} e^{ikx} = \frac{\sin(2\pi Nx/L)}{\pi x}.$$

So it is clear that we're defining the $\delta(x)$ as a limit of the integral $\int_{-2\pi N/L}^{2\pi N/L} (dk/2\pi)e^{ikx}$, which is abruptly cut off at the large values $\pm(2\pi N/L)$. In fact, this is not very physical: a much more realistic scenario for a *real* wave packet would be a *gradual diminution* in contributions from high frequency (or short wavelength) modes—that is to say, a *gentle* cutoff in the integral over k that was used to replace the sum over n. For example, a reasonable cutoff procedure would be to multiply the integrand by $\exp(-\Delta^2 k^2)$, then take the limit of small Δ.

Therefore a more reasonable definition of the delta function, from a physicist's point of view, would be

$$\delta(x) = \lim_{\Delta\to 0} \int_{-\infty}^{+\infty} \frac{dk}{2\pi} e^{ikx} e^{-\Delta^2 k^2} = \lim_{\Delta\to 0} \frac{1}{(4\pi\Delta^2)^{1/2}} e^{-x^2/4\Delta^2} = \lim_{\Delta\to 0} \delta_\Delta(x), \text{ say.}$$

That is to say, the delta function can be defined as the "narrow limit" of a Gaussian wave packet with total area 1. Unlike the function , has no oscillating sidebands, thanks to our smoothing out of the upper k-space cutoff, so step discontinuities do not generate Gibbs' phenomenon overshoot—instead, a step will be smoothed out over a distance of order .

PROPERTIES OF THE DELTA FUNCTION

It is straightforward to verify the following properties from the definition as a limit of a Gaussian wavepacket:

$$\int \delta(x)dx = 1, \quad \delta(x) = 0 \text{ for } x \neq 0.$$

$$\delta(x) = \delta(-x), \quad \delta(ax) = \frac{1}{|a|}\delta(x),$$

$$\int \delta(a-x)\delta(x-b)dx = \delta(a-b).$$

There is no unique way to define the delta function, and other cutoff procedures can give useful insights. For example, the k-space integral can be split into two and simple exponential cutoffs applied to the two halves, that is, we could take the definition to be

$$\delta(x) = \lim_{\varepsilon\to 0}\left(\int_{-\infty}^{0}\frac{dk}{2\pi}e^{ikx}e^{\varepsilon k} + \int_{0}^{+\infty}\frac{dk}{2\pi}e^{ikx}e^{-\varepsilon k}\right).$$

Evaluating the integrals,

$$\delta(x) = \lim_{\varepsilon\to 0}\frac{1}{2\pi}\left(\frac{1}{ix+\varepsilon}-\frac{1}{ix-\varepsilon}\right) = \lim_{\varepsilon\to 0}\frac{1}{\pi}\left(\frac{\varepsilon}{x^2+\varepsilon^2}\right).$$

It is easy to check that this function is correctly normalized by making the change of variable $x = \varepsilon\tan\theta$ and integrating from $-\pi/2$ to $\pi/2$. This representation of the delta function will prove to be useful later. Note that regarded as a function of a complex variable, the delta function has two poles on the pure imaginary axis at $z = \pm i\varepsilon$.

The standard definition of the principal value integral is:

$$\int_{-D}^{D} f(x)\frac{P}{x}dx = \lim_{\varepsilon\to 0}\left(\int_{-D}^{-\varepsilon}\frac{f(x)}{x}dx + \int_{\varepsilon}^{D}\frac{f(x)}{x}dx\right).$$

It is not difficult to see that for a continuous differentiable function $f(x)$ this is equivalent to

Therefore the principal value operator can be written symbolically:

$$\frac{P}{x} = \lim_{\varepsilon\to 0}\frac{x}{x^2+\varepsilon^2} = \lim_{\varepsilon\to 0}\frac{1}{2}\left(\frac{1}{x+i\varepsilon}+\frac{1}{x-i\varepsilon}\right).$$

Putting this together with the similar representation of the delta function above, and taking the limit of $\varepsilon\to 0$ to be understood, we have the useful result:

$$\frac{1}{x\pm i\varepsilon} = \frac{P}{x} \mp i\pi\delta(x).$$

Chapter 7

Linear Algebra for Quantum Mechanics

We've seen that in quantum mechanics, the state of an electron in some potential is given by a wave function $\psi(\vec{x},t)$, and physical variables are represented by operators on this wave function, such as the momentum in the x-direction $p_x = -i\hbar\partial/\partial x$. The Schrodinger wave equation is a *linear* equation, which means that if ψ_1 and ψ_2 are solutions, then so is $c_1\psi_1 + c_2\psi_2$, where c_1, c_2 are arbitrary complex numbers.

This linearity of the sets of possible solutions is true generally in quantum mechanics, as is the representation of physical variables by operators on the wave functions. The mathematical structure this describes, the linear set of possible states and sets of operators on those states, is in fact a *linear algebra* of operators acting on a *vector space*. From now on, this is the language we'll be using most of the time. To clarify, we'll give some definitions.

Vector Space

The prototypical vector space is of course the set of real vectors in ordinary three-dimensional space, these vectors can be represented by trios of real numbers (v_1, v_2, v_3) measuring the components in the x, y and z directions respectively.

The basic properties of these vectors are:

Any vector multiplied by a number is another vector in the space, $a(v_1, v_2, v_3) = (av_1, av_2, av_3)$;

the sum of two vectors is another vector in the space, that given by just adding the corresponding components together: $(v_1 + w_1, v_2 + w_2, v_3 + w_3)$.

These two properties together are referred to as "*closure*": adding vectors and multiplying them by numbers cannot get you out of the space.

A further property is that there is a unique null vector $(0,0,0)$ and each vector has an additive inverse $(-v_1, -v_2, -v_3)$ which added to the original vector gives the null vector.

Mathematicians have generalized the definition of a vector space: a general vector space has the properties we've listed above for three-dimensional real vectors, but the operations of addition and multiplication by a number are generalized to more abstract operations between more general entities. The operators are, however, restricted to being commutative and associative.

Notice that the list of necessary properties for a general vector space does not include that the vectors have a magnitude—that would be an additional requirement, giving what is called a *normed* vector space.

To go from the familiar three-dimensional vector space to the vector spaces relevant to quantum mechanics, first the real numbers (components of the vector and possible multiplying factors) are to be generalized to complex numbers, and second the three-component vector goes an n component vector. The consequent n-dimensional complex space is sufficient to describe the quantum mechanics of angular momentum, an important subject. But to describe the wave function of a particle in a box requires an *infinite* dimensional space, one dimension for each Fourier component, and to describe the wave function for a particle on an infinite line requires the set of all normalizable continuous differentiable functions on that line. Fortunately, all these generalizations are to finite or infinite sets of complex numbers, so the mathematicians' vector space requirements of commutativity and associativity are always trivially satisfied.

We use Dirac's notation for vectors, $|1\rangle, |2\rangle$ and call them "kets", so, in his language, if $|1\rangle, |2\rangle$ belong to the space, so does $c_1|1\rangle + c_2|2\rangle$ for arbitrary complex constants c_1, c_2. Since our vectors are made up of complex numbers, multiplying any vector by zero gives the null vector, and the additive inverse is given by reversing the signs of all the numbers in the vector.

Clearly, the set of solutions of Schrodinger's equation for an electron in a potential satisfies the requirements for a vector space: $\psi(\vec{x},t)$ is just a complex number at each point in space, so only complex numbers are involved in forming $c_1\psi_1 + c_2\psi_2$, and commutativity, associativity, etc., follow at once.

VECTOR SPACE DIMENSIONALITY

The vectors $|1\rangle, |2\rangle, |3\rangle$ are *linearly independent* if $c_1|1\rangle + c_2|2\rangle + c_3|3\rangle = 0$ implies $c_1 = c_2 = c_3 = 0$.

A vector space is n-dimensional if the maximum number of linearly independent vectors in the space is n.

Such a space is often called $V^n(C)$, or $V^n(R)$ if only real numbers are used.

Now, vector spaces with finite dimension n are clearly insufficient for describing functions of a continuous variable x. But they are well worth reviewing here: they are fine for describing quantized angular momentum, and they serve as a natural introduction to the infinite-dimensional spaces needed to describe spatial wavefunctions.

A set of n linearly independent vectors in n-dimensional space is a *basis*—any vector can be written *in a unique way* as a sum over a basis:

$$|V\rangle = \sum v_i |i\rangle$$

You can check the uniqueness by taking the difference between two supposedly distinct sums: it will be a linear relation between independent vectors, a contradiction.

Since all vectors in the space can be written as linear sums over the elements of the basis, the sum of multiples of any two vectors has the form:

$$a|V\rangle + b|W\rangle = \sum (av_i + bw_i)|i\rangle.$$

INNER PRODUCT SPACES

The vector spaces of relevance in quantum mechanics also have an operation associating a number with a pair of vectors, a generalization of the dot product of two ordinary three-dimensional vectors,

$$\vec{a}.\vec{b} = \sum a_i b_i.$$

Following Dirac, we write the inner product of two ket vectors $|V\rangle$, $|W\rangle$ as $\langle W|V\rangle$. Dirac refers to this $\langle\ |\ \rangle$ form as a "bracket" made up of a "bra" and a "ket".

An *important difference* from the simple three-dimensional real vector inner product is that we're usually dealing with *complex* vector spaces, and *we require*

$$\langle W|V\rangle = \langle V|W\rangle^*$$

where * denotes complex conjugate. This means that $\langle V|V\rangle$ is real. We require $\langle V|V\rangle$ to be *positive,* except for the null vector, for which it is zero. (These requirements do restrict the classes of vector spaces we are considering, but they are all satisfied by the spaces relevant to quantum mechanics.)

The *norm* of $|V\rangle$ is defined by

$$|V| = \sqrt{\langle V|V\rangle}.$$

If $|V\rangle$ is a member of $V^n(C)$, so is $a|V\rangle$, for any complex number a. We require the inner product operation to commute with multiplication by a number, so $\langle W|(a|V\rangle) = a\langle W|V\rangle$. The complex conjugate of the right hand side is $a^*\langle V|W\rangle$. For consistency, the bra corresponding to the ket $a|V\rangle$ must be $\langle V|a^*$.

It follows that if

$$|V\rangle = \sum v_i|i\rangle,\quad |W\rangle = \sum w_i|i\rangle,\ \text{then}\ \langle V|W\rangle = \sum v_i{}^* w_j\langle i|j\rangle.$$

CONSTRUCTING AN ORTHONORMAL BASIS: THE GRAM-SCHMIDT PROCESS

To have something better resembling the standard dot product of ordinary three vectors, we need $\langle i|j\rangle = \delta_{ij}$, that is, we need to construct an *orthonormal basis* in the space. There is a straightforward procedure for doing this called the Gram-Schmidt process. We begin with a linearly independent set of basis vectors, $|1\rangle, |2\rangle, |3\rangle, \ldots$

We first normalize $|1\rangle$ by dividing it by its norm. Call the normalized vector $|1\rangle$. Now $|2\rangle$ cannot be parallel to $|1\rangle$, because the original basis was of linearly independent vectors, but $|2\rangle$ in general has a nonzero component parallel to $|1\rangle$ equal to $|I\rangle\langle I|2\rangle$, since $|1\rangle$ is normalized. Therefore, the vector $|2\rangle - |I\rangle\langle I|2\rangle$ is perpendicular to $|1\rangle$ as is easily verified. It is also easy to compute the norm of this vector, and divide by it to get $|II\rangle$, the second member of the orthonormal basis. Next, we take $|3\rangle$ and subtract off its components in the directions $|1\rangle$ and $|II\rangle$, and normalize the remainder, and so on.

In an n-dimensional space, having constructed an orthonormal basis with members $|i\rangle$, any vector $|V\rangle$ can be written as a column vector,

The corresponding bra is $\langle V| = \sum v_i{}^*\langle i|$, which we write as a row vector with the elements complex conjugated,

$\langle V| = (v_1^*, v_2^*, \ldots v_n^*)$. This operation, going from columns to rows and taking the complex conjugate, is called taking the *adjoint,* and can also be applied to matrices, as we shall see shortly.

$$|V\rangle = \sum v_i |i\rangle = \begin{pmatrix} v_1 \\ v_2 \\ . \\ . \\ v_n \end{pmatrix}, \text{ where } |1\rangle = \begin{pmatrix} 1 \\ 0 \\ . \\ . \\ 0 \end{pmatrix} \text{ and so on.}$$

The reason for representing the bra as a row is that the inner product of two vectors is then given by standard matrix multiplication:

$$\langle V|W\rangle = (v_1^*, v_2^*, \ldots, v_n^*) \begin{pmatrix} w_1 \\ . \\ . \\ w_n \end{pmatrix}.$$

(Of course, this only works with an orthonormal base.)

THE SCHWARTZ INEQUALITY

The Schwartz inequality is the generalization to any inner product space of the result $|\vec{a}.\vec{b}|^2 \leq |\vec{a}|^2 |\vec{b}|^2$ (or $\cos^2\theta \leq 1$) for ordinary three-dimensional vectors. The equality sign in that result only holds when the vectors are parallel. To generalize to higher dimensions, one might just note that two vectors are in a two-dimensional subspace, but an illuminating way of understanding the inequality is to write the vector $\vec{a}$ as a sum of two components, one parallel to $\vec{b}$ and one perpendicular to . The component parallel $\vec{b}$ to is just $\vec{b}\left(\vec{a}\cdot\vec{b}\right)/\left|\vec{b}\right|^2$, so the component perpendicular to $\vec{b}$ is the vector $\vec{a}_\perp = \vec{a} - \vec{b}\left(\vec{a}\cdot\vec{b}\right)/\left|\vec{b}\right|^2$. Substituting this expression into $\vec{a}_\perp \cdot \vec{a}_\perp \geq 0$, the inequality follows.

This same point can be made in a general inner product space: if $|V\rangle, |W\rangle$ are two vectors, then

$$|Z\rangle = |V\rangle - \frac{|W\rangle\langle W|V\rangle}{|W|^2}$$

is the component of $|V\rangle$ perpendicular to $|W\rangle$, as is easily checked by taking its inner product with $|W\rangle$.

$\langle Z|Z\rangle \geq 0$ gives immediately $|\langle V|W\rangle|^2 \leq |V|^2 |W|^2$.

LINEAR OPERATORS

A *linear operator A* takes any vector in a linear vector space to a vector in that space, $A|V\rangle = |V'\rangle$, and satisfies

$$A\left(c_1|V_1\rangle + c_2|V_2\rangle\right) = c_1 A|V_1\rangle + c_2 A|V_2\rangle,$$

with c_1, c_2 arbitrary complex constants.

The *identity operator I* is (obviously!) defined by:

$I|V\rangle = |V\rangle$ for all $|V\rangle$.

For an n-dimensional vector space with an orthonormal basis $|1\rangle, \ldots, |n\rangle$, since any vector in the space can be expressed as a sum $|V\rangle = \sum v_i |i\rangle$, the linear operator is completely determined by its action on the basis vectors—this is all we need to know. It's easy to find an expression for the identity operator in terms of bras and kets.

Taking the inner product of both sides of the equation $|V\rangle = \sum v_i |i\rangle$ with the bra $\langle i|$ gives $\langle i|V\rangle = v_i$, so

$$|V\rangle = \sum v_i |i\rangle = \sum |i\rangle\langle i|V\rangle.$$

Since this is true for any vector in the space, it follows that that the identity operator is just

$$I = \sum_1^n |i\rangle\langle i|$$

This is an important result that will reappear in many disguises.

To analyse the action of a general linear operator A, we just need to know how it acts on each basis vector. Beginning with $A|1\rangle$, this must be some sum over the basis vectors, and since they are orthonormal, the component in the $|i\rangle$ direction must be just $\langle i|A|1\rangle$.

That is,

$$A|1\rangle = \sum_1^n |i\rangle\langle i|A|1\rangle = \sum_1^n A_{i1}|i\rangle, \text{ writing } \langle i|A|1\rangle = A_{i1}.$$

So if the linear operator A acting on $|V\rangle = \sum v_i|i\rangle$ gives $|V'\rangle = \sum v_i'|i\rangle$, that is, the $A|V\rangle = |V'\rangle$, linearity tells us that

$$\sum v_i'|i\rangle = |V'\rangle = A|V\rangle = \sum v_j A|j\rangle = \sum_{i,j} v_j|i\rangle\langle i|A|j\rangle = \sum_{i,j} v_j A_{ij}|i\rangle$$

where in the fourth step we just inserted the identity operator.

Since the $|i\rangle$'s are all orthogonal, the coefficient of a particular $|i\rangle$ on the left-hand side of the equation must be identical with the coefficient of the same $|i\rangle$ on the right-hand side. That is, $v_i' = A_{ij}v_j$.

Therefore the operator A is simply equivalent to *matrix multiplication*:

$$\begin{pmatrix} v_1' \\ v_2' \\ \cdot \\ \cdot \\ v_n' \end{pmatrix} = \begin{pmatrix} \langle 1|A|1\rangle & \langle 1|A|2\rangle & \cdot & \cdot & \langle 1|A|n\rangle \\ \langle 2|A|1\rangle & \langle 2|A|2\rangle & \cdot & \cdot & \cdot \\ \cdot & \cdot & \cdot & \cdot & \cdot \\ \cdot & \cdot & \cdot & \cdot & \cdot \\ \langle n|A|1\rangle & \langle n|A|2\rangle & \cdot & \cdot & \langle n|A|n\rangle \end{pmatrix} \begin{pmatrix} v_1 \\ v_2 \\ \cdot \\ \cdot \\ v_n \end{pmatrix}.$$

It is important to note that a linear operator applied successively to the members of an orthonormal basis might give a new set of vectors which *no longer span the entire space.* To give an example, the linear operator $|1\rangle\langle 1|$ applied to any vector in the space picks out the vector's component in the

$|1\rangle$ direction. It is called a *projection operator*. The operator $(|1\rangle\langle 1|+|2\rangle\langle 2|)$ projects a vector into its components in the subspace spanned by the vectors $|1\rangle$ and $|2\rangle$, and so on—if we extend the sum to be over the whole basis, we recover the identity operator.

THE ADJOINT OPERATOR AND HERMITIAN MATRICES

As we've discussed, if a ket $|V\rangle$ in the n-dimensional space is written as a column vector with n (complex) components, the corresponding bra is a *row* vector having as elements the complex conjugates of the ket elements. $\langle W|V\rangle = \langle V|W\rangle^*$ then follows automatically from standard matrix multiplication rules, and on multiplying $|V\rangle$ by a complex number a to get $a|V\rangle$ (meaning that each element in the column of numbers is multiplied by a) the corresponding bra goes to $\langle V|a^* = a^*\langle V|$.

But suppose that instead of multiplying a ket by a number, we operate on it with a linear operator. What generates the parallel transformation among the bras? In other words, if $A|V\rangle = |V'\rangle$, can we find a corresponding operator that sends the bra $\langle V|$ to $\langle V'|$? This must be another linear operator, because A is linear, that is, if under A $|V_1\rangle \to |V_1'\rangle, |V_2\rangle \to |V_2'\rangle$ and $|V_3\rangle = |V_1\rangle + |V_2\rangle$, then under A $|V_3\rangle$ is required to go to $|V_3'\rangle = |V_1'\rangle + |V_2'\rangle$. Consequently, under the parallel *bra* transformation we must have $\langle V_1| \to \langle V_1'|, \langle V_2| \to \langle V_2'|$ and $\langle V_3| \to \langle V_3'|$, so the bra transformation is necessarily also linear. Recalling that the bra is an n-element *row* vector, the most general linear transformation sending it to another bra is an $n \times n$ matrix operating on the bra from the *right*.

This bra operator is called the *adjoint* of A, written $A^\dagger$. That is, the ket $A|V\rangle$ has corresponding bra $\langle V|A^\dagger$. In an orthonormal basis, using the notation $\langle Ai|$ to denote the bra $\langle i|A^\dagger$ corresponding to the ket $A|i\rangle = |Ai\rangle$, say,

$$(A^\dagger)_{ij} = \langle i|A^\dagger|j\rangle = \langle Ai|j\rangle = \langle j|Ai\rangle^* = A_{ji}{}^*.$$

So *the adjoint operator is the transpose complex conjugate.*

for a *product* of two operators (prove this!), $(AB)^\dagger = B^\dagger A^\dagger$.

An operator equal to its adjoint $A = A^\dagger$ is called *Hermitian,* one equal to minus its adjoint $A = -A^\dagger$ is anti Hermitian (sometimes skew Hermitian). These operator types are in some ways generalizations of real and imaginary numbers. Any operator can be expressed as a sum of a Hermitian operator and an anti Hermitian operator, $A = \frac{1}{2}(A + A^\dagger) + \frac{1}{2}(A - A^\dagger)$.

This notion naturally extends to *vectors and numbers*: the adjoint of a ket is the corresponding bra, the adjoint of a number is its complex conjugate. This is useful to bear in mind when taking the adjoint of an operator which may be partially constructed of vectors and numbers, such as projection-type operators. The adjoint of a product of matrices, vectors and numbers is the product of the adjoints *in reverse order.* (Of course, for numbers the order doesn't matter.)

UNITARY OPERATORS

An operator is *unitary* if $U^\dagger U = 1$. This implies first that U operating on any vector gives a vector having the same norm, since the new norm $\langle V|U^\dagger U|V\rangle = \langle V|V\rangle$ and furthermore inner products are preserved, $\langle W|U^\dagger U|V\rangle = \langle W|V\rangle$. The original orthonormal basis in the space must therefore under a unitary transformation go to another orthonormal basis.

Conversely, any transformation that takes one orthonormal basis into another one is a unitary transformation. To see this, suppose that a linear transformation A sends the members of the orthonormal basis $(|1\rangle_1, |2\rangle_1, \ldots, |n\rangle_1)$ to the different orthonormal set $(|1\rangle_2, |2\rangle_2, \ldots, |n\rangle_2)$, so $A|1\rangle_1 = |1\rangle_2$, etc. Then the vector $|V\rangle = \sum v_i |i\rangle_1$ will go to $|V'\rangle = A|V\rangle = \sum v_i |i\rangle_2$, having the same norm, from $\langle V|V\rangle = \sum |v_i|^2$, and similarly $\langle W|V\rangle$ is invariant.

A unitary operation amounts to a rotation (possibly combined with a reflection) in the space. Evidently, since $U^{\dagger}U = 1$, the adjoint $U^{\dagger}$ rotates the basis back—it is the inverse operation, and so $UU^{\dagger} = 1$ also, that is, U and commute.

DETERMINANTS

We review in this section the *determinant* of a matrix, a function closely related to the operator properties of the matrix.

Let's start with 2×2 matrices:

$$A = \begin{pmatrix} a_{11} & a_{12} \\ a_{21} & a_{22} \end{pmatrix}.$$

The determinant of this matrix is defined by:

$$\det A = |A| = a_{11}a_{22} - a_{12}a_{21}.$$

Writing the two rows of the matrix as vectors:

$$\vec{a}_1^R = (a_{11}, a_{12})$$
$$\vec{a}_2^R = (a_{21}, a_{22})$$

(R denotes row), $\det A = \vec{a}_1^R \times \vec{a}_2^R$ is just the *area* (with appropriate sign) of the parallelogram having the two row vectors as adjacent sides:

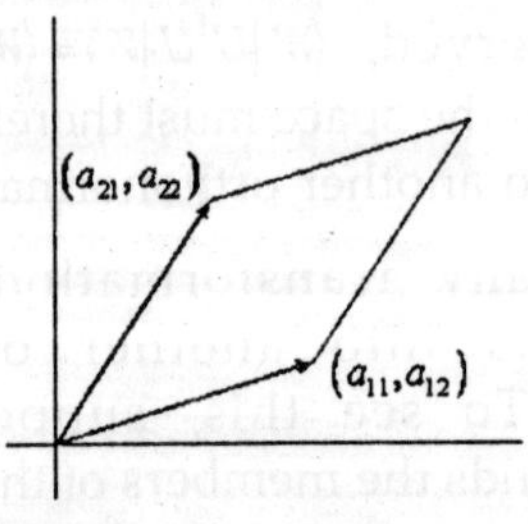

This is zero if the two vectors are parallel (linearly dependent) and is not changed by adding any multiple of $\vec{a}_2^R$ to $\vec{a}_1^R$ (because the new parallelogram has the same base and the same height as the original—check this by drawing).

Let's go on to the more interesting case of 3×3 matrices:

$$A=\begin{pmatrix} a_{11} & a_{12} & a_{13} \\ a_{21} & a_{22} & a_{23} \\ a_{31} & a_{32} & a_{33} \end{pmatrix}.$$

The determinant of A is defined as

$$\det A=\varepsilon_{ijk}a_{1i}a_{2j}a_{3k}$$

where $\varepsilon_{ijk}=0$ if any two are equal, +1 if ijk = 123, 231 or 312 (that is to say, an even permutation of 123) and –1 if ijk is an odd permutation of 123. Repeated suffixes, of course, imply summation here.

Writing this out explicitly,

$$\det A=a_{11}a_{22}a_{33}+a_{21}a_{32}a_{13}+a_{31}a_{12}a_{23}-a_{11}a_{32}a_{23}-a_{21}a_{12}a_{33}-a_{31}a_{22}a_{13}.$$

Just as in two dimensions, it's worth looking at this expression in terms of vectors representing the rows of the matrix

$$\vec{a}_1^R=(a_{11},a_{12},a_{13})$$
$$\vec{a}_2^R=(a_{21},a_{22},a_{23})$$
$$\vec{a}_3^R=(a_{31},a_{32},a_{33})$$

so

$$A=\begin{pmatrix} \vec{a}_1^R \\ \vec{a}_2^R \\ \vec{a}_3^R \end{pmatrix}, \quad \text{and we see that } \det A=(\vec{a}_1^R\times\vec{a}_2^R)\cdot\vec{a}_3^R.$$

This is the volume of the parallelopiped formed by the three vectors being adjacent sides (meeting at one corner, the origin).

This parallelepiped volume will of course be zero if the three vectors lie in a plane, and it is not changed if a multiple of one of the vectors is added to another of the vectors. That is to say, *the determinant of a matrix is not changed if a multiple of one row is added to another row.*

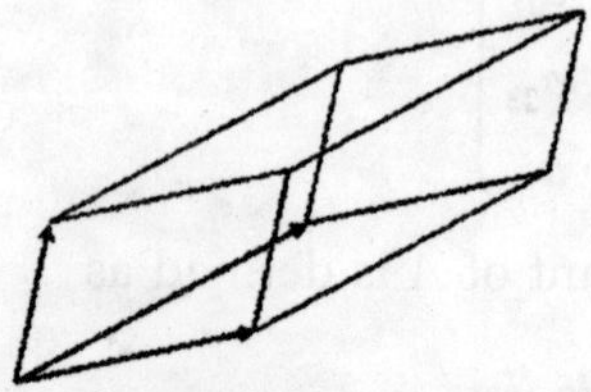

This is because the determinant is linear in the elements of a single row,

$$\det\begin{pmatrix}\vec{a}_1^R+\lambda\vec{a}_2^R\\ \vec{a}_2^R\\ \vec{a}_3^R\end{pmatrix}=\det\begin{pmatrix}\vec{a}_1^R\\ \vec{a}_2^R\\ \vec{a}_3^R\end{pmatrix}+\lambda\det\begin{pmatrix}\vec{a}_2^R\\ \vec{a}_2^R\\ \vec{a}_3^R\end{pmatrix}$$

and the last term is zero because two rows are identical—so the triple vector product vanishes.

A more general way of stating this, applicable to larger determinants, is that for a determinant with two identical rows, the symmetry of the two rows, together with the *antisymmetry* ϵ_{ijk}, of ensures that the terms in the sum all cancel in pairs.

Since the determinant is not altered by adding some multiple of one row to another, if the rows are linearly dependent, one row could be made identically zero by adding the right multiples of the other rows. Since every term in the expression for the determinant has one element from each row, the determinant would then be identically zero. For the three-dimensional case, the linear dependence of the rows means the corresponding vectors lie in a plane, and the parallelepiped is flat.

The algebraic argument generalizes easily to n×n determinants: they are *identically zero if the rows are linearly dependent.*

The generalization from 3×3 to $n\times n$ determinants is that $\det A=\varepsilon_{ijk}a_{1i}a_{2j}a_{3k}$ becomes:

$$\det A=\varepsilon_{ijk\ldots p}a_{1i}a_{2j}a_{3k}\ldots a_{np}$$

where $ijk...p$ is summed over all permutations of $132...n$, and the symbol is zero if any two of its suffixes are equal, +1 for an even permutation and -1 for an odd permutation. (*Note*: any permutation can be written as a product of swaps of neighbours. Such a representation is in general not unique, but for a given permutation, all such representations will have either an odd number of elements or an even number.)

An important theorem is that for a product of two matrices A, B the determinant of the product is the product of the determinants, $\det AB = \det A \times \det B$. This can be verified by brute force for 2×2 matrices, and a proof in the general case can be found in any book on mathematical physics.

It can also be proved that *if* the rows are linearly independent, the determinant cannot be zero.

(*Here's a proof*: take an matrix with the n row vectors linearly independent. Now consider the components of those vectors in the $n-1$ dimensional subspace perpendicular to $(1, 0, \ldots, 0)$. These n vectors, each with only $n-1$ components, must be linearly dependent, since there are more of them than the dimension of the space. So we can take some combination of the rows below the first row and subtract it from the first row to leave the first row $(a, 0, 0, \ldots, 0)$, and a cannot be zero since we have a matrix with n linearly independent rows. We can then subtract multiples of this first row from the other rows to get a determinant having zeros in the first column below the first row. Now look at the $n-1$ by $n-1$ determinant to be multiplied by a.

Its rows must be linearly independent since those of the original matrix were. Now proceed by induction.)

To return to three dimensions, it is clear from the form of

$$\det A = a_{11}a_{22}a_{33} + a_{21}a_{32}a_{13} + a_{31}a_{12}a_{23} - a_{11}a_{32}a_{23} - a_{21}a_{12}a_{33} - a_{31}a_{22}a_{13}$$

that we could equally have taken the *columns* of A as three vectors, $A = (\vec{a}_1^C, \vec{a}_2^C, \vec{a}_3^C)$ in an obvious notation, $\det A = (\vec{a}_1^C \times \vec{a}_2^C) \cdot \vec{a}_3^C$, and linear dependence among the *columns* will also ensure the vanishing of the determinant—

so, in fact, linear dependence of the columns ensures linear dependence of the rows.

This, too, generalizes to : in the definition of determinant det $A = \varepsilon_{ijk\ldots y} a_{1i} a_{2j} a_{3k} \ldots a_{ny}$, the row suffix is fixed and the column suffix goes over all permissible permutations, with the appropriate sign—but the same terms would be generated by having the *column* suffixes kept in numerical order and allowing the row suffix to undergo the permutations.

AN ASIDE: RECIPROCAL LATTICE VECTORS

It is perhaps worth mentioning how the inverse of a 3×3 matrix operator can be understood in terms of vectors. For a set of linearly independent vectors $(\vec{a}_1, \vec{a}_2, \vec{a}_3)$, a reciprocal set $(\vec{b}_1, \vec{b}_2, \vec{b}_3)$ can be defined by

$$\vec{b}_1 = \frac{\vec{a}_2 \times \vec{a}_3}{\vec{a}_1 \times \vec{a}_2 \cdot \vec{a}_3}$$

and the obvious cyclic definitions for the other two reciprocal vectors. We see immediately that

$$\vec{a}_i \cdot \vec{b}_j = \delta_{ij}$$

from which it follows that the inverse matrix to

$$A = \begin{pmatrix} \vec{a}_1^{\,R} \\ \vec{a}_2^{\,R} \\ \vec{a}_3^{\,R} \end{pmatrix} \text{ is } B = \begin{pmatrix} \vec{b}_1^{\,C} & \vec{b}_2^{\,C} & \vec{b}_3^{\,C} \end{pmatrix}.$$

(These reciprocal vectors are important in *x*-ray crystallography, for example. If a crystalline lattice has certain atoms at positions $n_1\vec{a}_1 + n_2\vec{a}_2 + n_3\vec{a}_3$, where n_1, n_2, n_3 are integers, the reciprocal vectors are the set of normals to possible planes of the atoms, and these planes of atoms are the important elements in the diffractive *x*-ray scattering.)

EIGENKETS AND EIGENVALUES

If an operator *A* operating on a ket $|V\rangle$ gives a multiple of the same ket,

$$A|V\rangle = \lambda|V\rangle$$

then $|V\rangle$ is said to be an *eigenket* (or, just as often, *eigenvector*, or *eigenstate*!) of A with *eigenvalue* .

Eigenkets and eigenvalues are of central importance in quantum mechanics: dynamical variables are operators, a physical measurement of a dynamical variable yields an eigenvalue of the operator, and forces the system into an eigenket.

In this section, we shall show how to find the eigenvalues and corresponding eigenkets for an operator A. We'll use the notation $A|a_i\rangle = a_i|a_i\rangle$ for the set of eigenkets $|a_i\rangle$ with corresponding eigenvalues a_i. (Obviously, in the eigenvalue equation here the suffix i not summed over.)

The first step in solving $A|V\rangle = \lambda|V\rangle$ is to find the allowed eigenvalues a_i.

Writing the equation in matrix form:

$$\begin{pmatrix} A_{11}-\lambda & A_{12} & . & . & A_{1n} \\ A_{21} & A_{22}-\lambda & . & . & . \\ . & . & . & . & . \\ . & . & . & . & . \\ A_{n1} & . & . & . & A_{nn}-\lambda \end{pmatrix} \begin{pmatrix} v_1 \\ v_2 \\ . \\ . \\ v_n \end{pmatrix} = 0.$$

This equation is actually telling us that the *columns* of the matrix $A - \lambda I$ are linearly dependent! To see this, write the matrix as a row vector each element of which is one of its columns, and the equation becomes

$$(\vec{M}_1^C, \vec{M}_2^C, \ldots, \vec{M}_n^C) \begin{pmatrix} v_1 \\ . \\ . \\ . \\ v_n \end{pmatrix} = 0$$

which is to say

$$v_1\vec{M}_1^C + v_2\vec{M}_2^C = \ldots + v_n\vec{M}_n^C = 0,$$

the columns of the matrix are indeed a *linearly dependent* set.

We know that means the determinant of the matrix $A-\lambda I$ is zero,

$$\begin{vmatrix} A_{11}-\lambda & A_{12} & . & . & A_{1n} \\ A_{21} & A_{22}-\lambda & . & . & . \\ . & . & . & . & . \\ . & . & . & . & . \\ A_{n1} & . & . & . & A_{nn}-\lambda \end{vmatrix} = 0.$$

Evaluating the determinant using $\det A = e_{jk\ldots p}a_{1i}a_{2j}a_{3k}\ldots a_{np}$ gives an nth order polynomial in sometimes called the *characteristic polynomial.* Any polynomial can be written in terms of its roots:

$$C(\lambda - a_1)(\lambda - a_2)\ldots(\lambda - a_n) = 0$$

where the a_i's, the roots of the polynomial, and C is an overall constant, which from inspection of the determinant we can see to be $(-1)^n$. (It's the coefficient of λ^n.) The polynomial roots (which we don't yet know) are in fact the eigenvalues. For example, putting $\lambda = a_1$ in the matrix, $\det(A - a_1 I) = 0$, which means that $(A - a_1 I)|V\rangle = 0$ has a nontrivial solution $|V\rangle$, and this is our eigenvector $|a_1\rangle$.

Notice that the diagonal term in the determinant $(A_{11}-\lambda)(A_{22}-\lambda)\ldots(A_{nn}-\lambda)$ generates the leading two orders in the polynomial $(-1)^n\left(\lambda^n - (A_{11}+\ldots+A_{nn})\lambda^{n-1}\right)$, (and some lower order terms too). Equating the coefficient of λ^{n-1} here with that in $(-1)^n(\lambda - a_1)(\lambda - a_2)\ldots(\lambda - a_n)$,

$$\sum_{i=1}^{n} a_i = \sum_{i=1}^{n} A_{ii} = \operatorname{Tr} A.$$

Putting $\lambda = 0$ in both the determinantal and the polynomial representations (in other words, equating the λ-independent terms),

$$\prod_{i=1}^{n} a_i = \det A.$$

So we can learn something about the eigenvalues directly from the determinant. But if the space is bigger than 2×2 this is not enough to solve the problem.

We must solve the polynomial equation $\det(A - \lambda I) = 0$ to find the set of eigenvalues, then use them to calculate the corresponding eigenvectors. This is done one at a time.

Labeling the first eigenvalue found as a_1, the corresponding equation for the components v_i of the eigenvector $|a_1\rangle$ is

$$\begin{pmatrix} A_{11}-a_1 & A_{12} & \cdot & \cdot & A_{1n} \\ A_{21} & A_{22}-a_1 & \cdot & \cdot & \cdot \\ \cdot & \cdot & \cdot & \cdot & \cdot \\ \cdot & \cdot & \cdot & \cdot & \cdot \\ A_{n1} & \cdot & \cdot & \cdot & A_{nn}-a_1 \end{pmatrix} \begin{pmatrix} v_1 \\ v_2 \\ \cdot \\ \cdot \\ v_n \end{pmatrix} = 0.$$

This looks like n equations for the n numbers v_i, but it isn't: remember the rows are linearly dependent, so there are only $n - 1$ independent equations. However, that's enough to determine

the ratios of the vector components $v_1, \ldots, v_n$, then finally the eigenvector is normalized.

EIGENVALUES AND EIGENSTATES OF HERMITIAN MATRICES

For a *Hermitian* matrix, it is easy to establish that the eigenvalues are always *real*. Taking (in this section) A to be hermitian, $A = A^\dagger$, and labeling the eigenkets by the eigenvalue, that is,

$$A|a_1\rangle = a_1|a_1\rangle$$

the inner product with the bra $\langle a_1|$ gives $\langle a_1|A|a_1\rangle = a_1\langle a_1|a_1\rangle$. But the inner product of the *adjoint* equation (remembering $A = A^\dagger$)

$$\langle a_1|A = a_1{}^*\langle a_1|$$

with $|a_1\rangle$ gives $\langle a_1|A|a_1\rangle = a_1{}^*\langle a_1|a_1\rangle$, so $a_1 = a_1{}^*$, and all the eigenvalues must be real.

Furthermore, the eigenkets belonging to *different* eigenvalues are *orthogonal,* if

$$A|a_1\rangle = a_1|a_1\rangle$$

$$A|a_2\rangle = a_2|a_2\rangle$$

taking the adjoint of the first equation and taking its inner product with $|a_2\rangle$,

$$\langle a_1|A|a_2\rangle = a_1\langle a_1|a_2\rangle = a_2\langle a_1|a_2\rangle$$

so $\langle a_1|a_2\rangle = 0$ unless the eigenvalues are equal. If they *are* equal, then any linear combination of them is also an eigenvector with the same eigenvalue, and the Gram Schmidt procedure can be used to construct an orthonormal basis in this subspace as part of a full orthonormal basis.

A subspace spanned by eigenvectors having the same eigenvalue is termed *degenerate.*

DIAGONALIZING A HERMITIAN MATRIX

Diagonalizing a matrix means transforming to an orthonormal basis in which the only nonzero matrix elements are on the diagonal, $i = j$.

In fact, it is evident from the discussion at the end of the previous section that a hermitian matrix is diagonal in the orthonormal basis of its set of eigenvectors: $|a_1\rangle, |a_2\rangle, \ldots, |a_n\rangle$, since

$$\langle a_i|A|a_j\rangle = \langle a_i|a_j|a_j\rangle = a_j\langle a_i|a_j\rangle = a_j\delta_{ij}.$$

However, we are assuming that we know the matrix elements of A in some other orthonormal basis, so to diagonalize A we need to rotate from the initial orthonormal basis to one made up of the eigenkets of A.

$$|1\rangle = \begin{pmatrix} 1 \\ 0 \\ 0 \\ . \\ 0 \end{pmatrix}, \quad |2\rangle = \begin{pmatrix} 0 \\ 1 \\ 0 \\ . \\ 0 \end{pmatrix}, \quad |i\rangle = \begin{pmatrix} 0 \\ \vdots \\ 1 \\ \vdots \\ 0 \end{pmatrix} \ldots (\bar{1} \text{ in } i^{\text{th}} \text{ place down}), \quad |n\rangle = \begin{pmatrix} 0 \\ 0 \\ 0 \\ . \\ 1 \end{pmatrix}.$$

Denoting the initial orthonormal basis in the standard fashion.

The elements of the matrix are $A_{ij} = \langle i|A|j\rangle$.

A transformation from one orthonormal basis to another is a *unitary* transformation, as discussed above, so we write it

$$|V\rangle \to |V'\rangle = U|V\rangle.$$

Under this transformation, the matrix element

$$\langle W|A|V\rangle \to \langle W'|A|V'\rangle = \langle W|U^{\dagger}AU|V\rangle.$$

So we can find the appropriate transformation matrix U by requiring that $U^{\dagger}AU$ be diagonal with respect to the *original* set of basis vectors. (Transforming the operator in this way, leaving the vector space alone, is equivalent to rotating the vector space and leaving the operator alone. Of course, in a system with more than one operator, the same transformation would have to be applied to all the operators).

In fact, the unitary matrix U we need is just composed of the normalized eigenkets of the operator A,

$$U = (|a_1\rangle, |a_2\rangle, \ldots, |a_n\rangle).$$

It is easy to see that this must be true:

$$AU = (a_1|a_1\rangle, a_2|a_2\rangle, \ldots, a_n|a_n\rangle)$$

so

$$\left(U^{\dagger}AU\right)_{ij} = \langle a_i | a_j | a_j \rangle = \delta_{ij} a_j, \text{ a diagonal matrix.}$$

(The repeated suffixes here are of course *not* summed over.)

To actually find the eigenkets, then, we need to go through the routine outlined in the previous section, that is, first find the eigen*values* by finding the zeros of the determinantal equation, $|A - \lambda I| = 0$, then enter the roots $a_1, a_2, \ldots, a_n$ one at a time into the matrix equation

$$\begin{pmatrix} A_{11} - a_i & A_{12} & . & . & A_{1n} \\ A_{21} & A_{22} - a_i & . & . & . \\ . & . & . & . & . \\ . & . & . & . & . \\ A_{n1} & . & . & . & A_{nn} - a_i \end{pmatrix} \begin{pmatrix} v_1 \\ v_2 \\ . \\ . \\ v_n \end{pmatrix} = 0$$

to give n - 1 linear equations for the n unknown components of the eigenket $|a_i\rangle$.

Then

$$U^{\dagger}AU = \begin{pmatrix} a_1 & 0 & 0 & . & 0 \\ 0 & a_2 & 0 & . & 0 \\ 0 & 0 & a_3 & . & 0 \\ . & . & . & . & . \\ 0 & . & . & . & a_n \end{pmatrix}$$

If some of the eigenvalues are the same, the Gram Schmidt procedure may be needed to generate an orthogonal set, as mentioned earlier.

FUNCTIONS OF MATRICES

The same unitary operator U that diagonalizes an Hermitian matrix A will also diagonalize A^2, because

$$U^{-1}A^2U = U^{-1}AAU = U^{-1}AUU^{-1}AU$$

so

$$U^{\dagger}A^{2}U=\begin{pmatrix} a_1^2 & 0 & 0 & . & 0 \\ 0 & a_2^2 & 0 & . & 0 \\ 0 & 0 & a_3^2 & . & 0 \\ . & . & . & . & . \\ 0 & . & . & . & a_n^2 \end{pmatrix}$$

Evidently, this same process works for any power of *A*, and formally for any function of *A* expressible as a power series, but of course convergence properties need to be considered, and this becomes trickier on going from finite matrices to operators on infinite spaces.

COMMUTING HERMITIAN MATRICES

From the above, the set of powers of an Hermitian matrix all commute with each other, and have a common set of eigenvectors (but not the same eigenvalues, obviously). In fact it is not difficult to show that any two Hermitian matrices that commute with each other have the same set of eigenvectors (after possible Gram Schmidt rearrangements in degenerate subspaces).

If two $n \times n$ Hermitian matrices *A*, *B* commute, that is, *AB* = *BA*, and *A* has a *nondegenerate* set of eigenvectors $A|a_i\rangle = a_i|a_i\rangle$, then $AB|a_i\rangle = BA|a_i\rangle = Ba_i|a_i\rangle = a_iB|a_i\rangle$, that is, $B|a_i\rangle$ is an eigenvector of *A* with eigenvalue a_i. Since *A* is nondegenerate $B|a_i\rangle$, must be some multiple of $|a_i\rangle$, and we conclude that *A*, *B* have the same set of eigenvectors.

Now suppose *A is* degenerate, and consider the $m \times m$ subspace S_{a_i} spanned by the eigenvectors $|a_i, 1\rangle, |a_i, 2\rangle, \ldots$ of *A* having eigenvalue a_i. Applying the argument in the paragraph above, $B|a_i, 1\rangle, B|a_i, 2\rangle, \ldots$ must also lie in this subspace. Therefore, if we transform *B* with the same unitary transformation that diagonalized *A*, *B* will not in general be diagonal in the subspace S_{a_i}, but it will be what is termed *block diagonal,* in that if B operates on any vector in S_{a_i} it gives a vector in S_{a_i}.

B can be written as two diagonal blocks: one $m \times m$, one $(n-m)\times(n-m)$with zeroes outside these diagonal blocks, for example, for $m = 2$, $n = 5$:

$$\begin{pmatrix} b_{11} & b_{12} & 0 & 0 & 0 \\ b_{21} & b_{22} & 0 & 0 & 0 \\ 0 & 0 & b_{33} & b_{34} & b_{35} \\ 0 & 0 & b_{43} & b_{44} & b_{45} \\ 0 & 0 & b_{53} & b_{54} & b_{55} \end{pmatrix}.$$

And, in fact, if there is only *one* degenerate eigenvalue that second block will only have nonzero terms on the diagonal:

$$\begin{pmatrix} b_{11} & b_{12} & 0 & 0 & 0 \\ b_{21} & b_{22} & 0 & 0 & 0 \\ 0 & 0 & b_{3} & 0 & 0 \\ 0 & 0 & 0 & b_{4} & 0 \\ 0 & 0 & 0 & 0 & b_{5} \end{pmatrix}.$$

B therefore operates on two subspaces, one m-dimensional, one $(n-m)$-dimensional, *independently*—a vector entirely in one subspace stays there.

This means we can complete the diagonalization of B with a unitary operator that *only* operates on the $m \times m$ block S_{a_i}. Such an operator will also affect the eigenvectors of A, but that doesn't matter, because all vectors in this subspace are eigenvectors of A with the *same* eigenvalue, so as far as A is concerned, we can choose any orthonormal basis we like—the basis vectors will still be eigenvectors

This establishes that any two *commuting* Hermitian matrices can be diagonalized at the same time. Obviously, this can never be true of *noncommuting* matrices, since all diagonal matrices commute.

DIAGONALIZING A UNITARY MATRIX

As previously stated, a unitary matrix is an operator that rotates an orthonormal basis into another orthonormal basis. It is clear that in three dimensions the axis through the origin about which the basis is rotated is an eigenvector of this transformation. It's less clear what the other two eigenvectors might be—or, equivalently, what are the eigenvectors corresponding to a two dimensional rotation of basis in a plane? The way to find out is to write down the matrix and diagonalize it.

The matrix

$$U(\theta) = \begin{pmatrix} \cos\theta & \sin\theta \\ -\sin\theta & \cos\theta \end{pmatrix}.$$

Note that the determinant is equal to unity. The eigenvalues are given by solving

$$\begin{vmatrix} \cos\theta - \lambda & \sin\theta \\ -\sin\theta & \cos\theta - \lambda \end{vmatrix} = 0 \text{ to give } \lambda = e^{\pm i\theta}$$

The corresponding eigenvectors satisfy

$$\begin{pmatrix} \cos\theta & \sin\theta \\ -\sin\theta & \cos\theta \end{pmatrix} \begin{pmatrix} u_1^{\pm} \\ u_2^{\pm} \end{pmatrix} = e^{\pm i\theta} \begin{pmatrix} u_1^{\pm} \\ u_2^{\pm} \end{pmatrix}.$$

The eigenvectors, normalized, are:

$$\begin{pmatrix} u_1^{\pm} \\ u_2^{\pm} \end{pmatrix} = \frac{1}{\sqrt{2}} \begin{pmatrix} 1 \\ \pm i \end{pmatrix}.$$

Note that, in contrast to a Hermitian matrix, the eigenvalues of a unitary matrix do not have to be real. In fact, from $U^{\dagger}U = 1$, sandwiched between the bra and ket of an eigenvector, we see that any eigenvalue of a unitary matrix must have unit modulus—it's a complex number on the unit circle. With hindsight, we should have realised that one eigenvalue of a two-dimensional rotation had to be $e^{i\theta}$, the product of two two-dimensional rotations is given be adding the angles of rotation, and a rotation through π changes all signs, so has eigenvalue

−1. Note that the eigenvector itself is independent of the angle of rotation—the rotations all commute, so they must have common eigenvectors. Successive rotation operators applied to the plus eigenvector add their angles, when applied to the minus eigenvector, all angles are subtracted.

Chapter 8

Function Spaces

LINEAR ALGEBRA IN INFINITE DIMENSIONS

The motivation for our review of linear algebra was the observation that the set of solutions to Schrodinger's equation satisfies some of the basic requirements of a vector space, in that linear combinations of solutions give another solution to the equation. Furthermore, Schrodinger's equation itself, as a differential operator acting on a function, suggests that the concept of a matrix operator acting on vectors in an n-dimensional vector space can be extended to more general operators, such as differential operators, acting on functions in an infinite-dimensional space.

Our analysis of linear vector spaces began by defining an inner product, which was used to establish an orthonormal basis for the space. Constructing a well-defined basis for the space of all functions on the real axis sounds impossible, and probably is. Fortunately, we don't need to be so all-encompassing. For one thing, we are not interested in functions with discontinuities, because in quantum mechanics that would be a wavefunction corresponding to infinite energy. (We can allow discontinuities in slope, although, as discussed in the Electron in a Box lecture, that occurs only where the potential is infinite. Infinite potentials are of course unphysical, but are convenient approximations in some cases, so we'll keep that option open.) Another important restriction arises from the requirement that the wavefunction describe a single particle—it must be normalizable, that is to say the norm.

$$\int_{-\infty}^{\infty} \psi^*(x)\psi(x)dx < \infty$$

and in fact ψ must be scaled so that this integral is equal to unity for actual computation of probabilities. Note that $\psi(x)$ means $\psi(x, t=0)$, but the norm turns out to be time-independent, as it must be, for the case of a single particle.

Building on the analogy with n-dimensional vector spaces, the requirement of finite norm suggests a definition for the inner product in function space:

$$\langle f|g\rangle = \int_{-\infty}^{\infty} f^*(x)g(x)dx.$$

This definition satisfies Dirac's requirement that $\langle f|g\rangle^* = \langle g|f\rangle$, gives a positive norm, and is linear in f, g. The space of functions with this inner product, and with finite norm $\sqrt{\langle f|f\rangle}$, is written $L_2(-\infty,\infty)$ or just L2. The functions are said to be "square integrable".

Notice that this inner product resembles the linear algebraic bra-ket product if we imagine every point on the line as an independent basis vector—mathematically meaningless, of course, but a hint of where we're going.

ELECTRON IN A BOX AGAIN

As a preliminary to discussing functions on the infinite line, it is worth considering those restricted to the finite interval (0, L) and vanishing at the two ends. These are precisely the conditions satisfied by the electron-in-a-box wavefunctions (see the earlier lecture):

$$|n\rangle = \psi_n(x,t=0) = \sqrt{\frac{2}{L}}\sin\frac{n\pi x}{L}.$$

Recall from the Fourier Series lecture that any function without discontinuities can be represented as a sum over Fourier components. For the present case of functions equal to zero at the two ends (as any physical wavefunction in a box must be) the sine kets $|n\rangle$ above form a complete set, that is to say, at t = 0, any $\psi(x)$ satisfying the boundary condition can be written

$$|\psi(x)\rangle = \sum_{n=1}^{\infty} a_n |n\rangle$$

where, from the orthonormality of the basis set $|n\rangle$, the Fourier coefficients $a_n = \langle n|\psi\rangle$, so (making explicit that $\psi(x)$ is in fact a ket in this vector space)

$$|\psi\rangle = \sum_{n=1}^{\infty} |n\rangle\langle n|\psi\rangle,$$

giving an identity operator in the space of continuous functions vanishing at 0 and L:

$$I = \sum_{n=1}^{\infty} |n\rangle\langle n|$$

exactly analogous to that in finite-dimensional vector spaces. The inner product of two functions

$$\psi(x) = \sum_{n=1}^{\infty} a_n |n\rangle, \quad \phi(x) = \sum_{n=1}^{\infty} b_n |n\rangle$$

defined as in the preceding section by

$$\langle\phi|\psi\rangle = \int_{-\infty}^{\infty} \phi^*(x)\psi(x)dx$$

is equivalently, in terms of Fourier coefficients,

$$\langle\phi|\psi\rangle = \sum_{n=1}^{\infty} b_n{}^* a_n$$

and the normalization

$$\langle\psi|\psi\rangle = \int_{-\infty}^{\infty} \psi^*(x)\psi(x)dx = \sum_{n=1}^{\infty} |a_n|^2 = 1 .$$

So for the electron-in-a-box wavefunctions, the orthonormal basis of sine functions gives a well-defined infinite-dimensional vector space.

We have previously stated that the standard interpretation of the wavefunction $\psi(x)$ is that $|\psi(x)|^2 dx$ is the probability of finding the particle in a small interval dx near x, and on integrating over all x the total probability of finding the particle is one. But we could also look for the particle in a particular state, rather than in a particular small interval dx. In this case, $|a_n|^2$ is the probability of finding the particle in the nth state. This is consistent with the previous interpretation, and is parallel to our earlier analysis of the probability of a particle having a particular momentum. The state coefficient an is called the amplitude, or sometimes the probability amplitude.

You might be wondering how we would measure that a particle is in a particular state. The answer is to wait for it to jump out. If an atom is excited (for example by a short burst of radiation) it will be excited to a state which is a linear superposition of different energy eigenstates, $\sum a_n |E_n\rangle$, rather than to a single eigenstate. It will usually return to the ground state by emitting one or a series of photons, and the frequency of an emitted photon reveals the energy difference between the atomic states involved. For a collection of atoms excited in the same way, the relative intensities of different spectral lines give the relative probabilities of different states. Of course, a long almost monochromatic wave packet of incoming radiation will tend to put all the excited atoms into the same state.

Exercise: Write out the identity operator for the electron in a box $I = \sum_{n=1}^{\infty} |n\rangle\langle n|$ using the explicit form $|n\rangle = \sqrt{\frac{2}{L}} \sin \frac{n\pi x}{L}$. Prove this is equivalent to the delta function when operating on other functions within the box. What is the behaviour of this function outside the box?

Functions on the Infinite Line

What happens if we take the analysis of the previous section and let L go to infinity? This is parallel to the analysis (two lectures back) of going from Fourier series to the Fourier transform, the sum over a series of plane waves satisfying a boundary condition becoming an integral over the continuum of all plane waves. In that lecture, we saw that as L went to infinity, the individual coefficients an went to zero as 1/L, but their density in momentum space increased as L, so the sum tended to a finite integral, specifically

$$\psi(x) = \frac{1}{2\pi}\int_{-\infty}^{\infty} a(k)e^{ikx}dk \text{ with } a(k) = \int_{-\infty}^{\infty} \psi(x)e^{-ikx}dx \ .$$

For the electron in a box (Fourier series) above we wrote the corresponding equation in Dirac notation as

$$|\psi(x)\rangle = \sum_{n=1}^{\infty} a_n|n\rangle \text{ with } a_n = \langle n|\psi\rangle, \text{ so } I = \sum_{n=1}^{\infty}|n\rangle\langle n|.$$

It's tempting to write down the analogous equations for the infinite line case, by translating the Fourier transform equations into Dirac notation:

$$|\psi(x)\rangle = \int_{-\infty}^{\infty}\frac{dk}{2\pi}a(k)|k\rangle,\ a(k) = \int_{-\infty}^{\infty}\psi(x)e^{-ikx}dx = \langle k|\psi(x)\rangle,\ \ I = \int_{-\infty}^{\infty}\frac{dk}{2\pi}|k\rangle\langle k|.$$

This looks good, but has a problem—in contrast to the Fourier series basis functions $|n\rangle$, these Fourier transform "basis states" $|k\rangle$ are infinitely long plane wave states e^{ikx} and therefore not normalizable in the sense we've used that term so far. They're not even in the space we're supposed to be working in!

Furthermore, $\langle k|\psi(x)\rangle$ is not the probability that a measurement of the momentum of the electron will yield precisely the value $p = \hbar k$. The correct probabilistic interpretation for a continuum of k-values is exactly parallel to the continuum of x-values in ordinary space:

$|\langle k|\psi(x)\rangle|^2 dk$ is the probability that a measurement of momentum would find the k-value to be in a small interval of width dk near k. The probability goes to zero with the width of the interval, and so is vanishingly small if we demand an exact value of k.

But we never measure k with infinite precision anyway—that would take an infinitely large apparatus. The physically significant quantity is the probability of finding k in a small interval dk—in practice, with real detectors, we are always integrating over some (small) range in k.

This means we might be ok with this continuum basis of states: we don't want them to be normalized in the traditional fashion $\langle k|k\rangle = 1$, because that would correspond to a finite probability of the particle having a mathematically precise value of k, which makes no physical sense—in fact it's nonsense. The normalization we need is one that makes sense in the context of an integral over a small interval in k—but still of course over a continuous infinity of basis states!

From our earlier definition of the delta function, we can express orthogonality of these $|k\rangle$ states:

$$\langle k'|k\rangle = \int_{-\infty}^{\infty} e^{i(k-k')x} dx = 2\pi\delta(k-k')$$

and, since the δ-function is normalized in the sense that it has total weight one in an integral, we take this equation as the definition of the normalization of the functions $|k\rangle$. That is to say, we take the state $|k\rangle$ to have wave function Ae^{ikx} with A = 1.

Exercise: from the expression for the identity operator above, $|\psi\rangle = \int_{-\infty}^{\infty} \frac{dk}{2\pi} |k\rangle\langle k|\psi\rangle$. Substitute $|k\rangle = e^{ikx}$ and check that this makes sense.

Note: some authors prefer to define the normalized plane wave states by $|k\rangle = \sqrt{1/2\pi} e^{ikx}$, in which case

$\langle k'|k\rangle = \delta(k'-k)$, and the $dk/2\pi$ appearing in the above integral for the identity operator becomes simply dk. With our convention, dk always appears with a 2π in the denominator.

Further Note: some prefer to go to a huge but not infinite box, so the basis momentum eigenstates wave functions are the discrete set $|k\rangle = \sqrt{\frac{1}{L}}e^{ikx}$, or in three dimensions $|k\rangle = \sqrt{\frac{1}{V}}e^{i\vec{k}\cdot\vec{r}}$ V being the volume. For this huge box, it is safe to replace the sum over discrete momentum states by an integral, bearing in mind that the density of states in phase space being proportional to L gives $\sum_{n} \equiv \int L\frac{dk}{2\pi}$ or $\int V\frac{d^3k}{(2\pi)^3}$ in three dimensions. The L or V factors finally cancel in computations.

Schrodinger's Equation as an Operator on a Vector Space

As we recounted at the beginning of this course, when Schrodinger was challenged to find a wave equation for the electron wave, he constructed one parallel to the electromagnetic "photon wave equation", that is to say, he took the energy-momentum equation and wrote

$$E = i\hbar\frac{\partial}{\partial t}, \quad p_x = -i\hbar\frac{\partial}{\partial x}.$$

He discovered that the three-dimensional version of the differential equation constructed in this way could be solved by standard analytic methods for an electron in an inverse-square potential—the hydrogen atom. The standing wave solutions yielded the right set of energy levels—those Bohr had found earlier with his simplistic model. This confirmed that indeed the wave equation describing the propagation of the electron waves had been discovered, and it was

$$\left(E - \frac{p^2}{2m} - V(x)\right)\psi(x,t) = 0$$

with E, p the differential operators given above. Since the operator in brackets is linear, the solutions $\psi(x,t)$ form a linear vector space.

Differential Operators: the Momentum Operator on L2

Our task now is to recast this old approach of differential operators acting on wave functions in the equivalent Dirac language. Let's begin with the simplest, the momentum operator. First, we need to show that it is Hermitian:

$$\langle\phi|p_x|\psi\rangle = -i\hbar\int_{-\infty}^{\infty} dx\,\phi^*(x)\frac{d\psi(x)}{dx} = i\hbar\int_{-\infty}^{\infty} dx\,\psi(x)\frac{d\phi^*(x)}{dx} - [\phi^*(x)\psi(x)]_{-\infty}^{\infty}.$$

The important point here is that the last term, the contribution from the infinite endpoints of the integration, must be zero because square-integrable functions must go to zero at infinity. Then it becomes apparent that

$$i\hbar\int_{-\infty}^{\infty} dx\,\psi(x)\frac{d\phi^*(x)}{dx} = \langle p_x\phi|\psi\rangle$$

with the appropriate sign change i going to ?i on taking the adjoint of $|p_x\phi\rangle$. (Another way to see what we have done here is to take the complex conjugate (or adjoint) of both sides of this last equation, yielding $\langle\phi|p_x|\psi\rangle^* = \langle\psi|p_x|\phi\rangle$.) The i is important: the differential operator d/dx alone is not Hermitian, it's anti Hermitian in L2!

So px is a Hermitian operator, and therefore has real eigenvalues, which it must have since momentum is a physical quantity. But what are its eigenvectors? We already know, of course, that they are the plane wave states—this is the whole reason this particular operator was chosen in building the wave equation in the first place. Strictly speaking, though, as we've already discussed, these plane wave states are not in L2. Nevertheless, any smooth function in L2 can be expressed as an integral over these states, so they do form a complete basis for the functions relevant to physics.

THE POSITION OPERATOR AND ITS EIGENSTATES

The "position" is just the co-ordinate x, manifestly always real, and therefore a Hermitian operator. We shall make clear that in this context we regard it as an operator by writing it with a little hat, $\hat{x}$. It is equally clear that the eigenstates of $\hat{x}$, states in which the particle has probability equal to one of being at a particular position, must be delta functions corresponding to that position: that's the only function with zero probability of finding the particle anywhere else. So if $|a\rangle$ is an eigenstate of x with eigenvalue a,

$$|a\rangle = C\delta(x-a) \quad (?)$$

where C is a constant. But, whatever value we choose for C, this wave function, like the momentum eigenstate, isn't normalizable—so, in fact, $|a\rangle$ itself could never be the wave function of a particle! (It wouldn't be physically reasonable anyway—to localize a particle to a point would take infinite energy.) But the set of all $|a\rangle$'s is certainly complete, and therein lies its value: it is a basis for the space. The convention is to "normalize" these kets, or rather to construct an "orthonormal set" by analogy with the orthonormalization convention for the plane-wave momentum states, that is, to take.

$$\langle a|b\rangle = \delta(a-b).$$

From the earlier result

$$\int \delta(a-x)\delta(x-b)dx = \delta(a-b)$$

it follows immediately that C = 1

Therefore,

$$\langle x|\psi\rangle = \int_{-\infty}^{\infty} dx'\delta(x-x')\psi(x') = \psi(x).$$

Taking the inner product of $|\psi\rangle$ with the bra $\langle x|$ just gives the value of ψ at the point x. Consequently any function $\psi(x)$ in L2 can be written:

$$|\psi\rangle=\int_{-\infty}^{\infty}dx|x\rangle\langle x|\psi\rangle$$

Exercise: check that this is true by finding $\langle x'|\psi\rangle$.

It follows from the above equation that the identity operator in L2 can be written in terms of the eigenstates of $\hat{x}$:

$$I=\int_{-\infty}^{\infty}dx|x\rangle\langle x|.$$

From this, $|k\rangle$ can be written

$$|k\rangle=\int_{-\infty}^{\infty}dx|x\rangle\langle x|k\rangle=\int_{-\infty}^{\infty}dxe^{ikx}|x\rangle,$$

and

$$|x\rangle=\int_{-\infty}^{\infty}\frac{dk}{2\pi}|k\rangle\langle k|x\rangle=\int_{-\infty}^{\infty}\frac{dk}{2\pi}e^{-ikx}|k\rangle.$$

These are possibly the least rigorous equations in this section—we're expressing one set of states outside of L2 in terms of another such set, using both sets as bases in L2! Obviously, this is only meaningful with an $|x\rangle$ state defined as a zero-width limit of narrowing Gaussians (say) and a $|k\rangle$ state as a limit of longer and longer wavepackets, tending to a single k-value. Yet despite the lack of rigor in the above presentation, these states, used with care, are in fact reliable and efficient tools for analyzing quantum mechanical problems. We shall use them often.

Exercise: show these equations are consistent by substituting $|k\rangle$ from the first into the right-hand side of the second, to give $|x\rangle=|x\rangle$.

The Hamiltonian Operator

The Hamiltonian operator gives the time development of the wavefunction. It corresponds to the total energy. If the wavefunction corresponds to a definite energy, the time dependence can be factored out, and the spatial wave function is a solution of Schrodinger's time independent equation:

$$H\psi(x)=\left(\frac{p^2}{2m}+V(x)\right)\psi(x)=-\frac{\hbar^2}{2m}\frac{\partial^2\psi(x)}{\partial x^2}+V(x)\psi(x)=E\psi(x).$$

Since we only consider the space L2 of wave functions on which both p and x are Hermitian, H must be Hermitian, and therefore has real eigenvalues.

The Basic Rules of Quantum Mechanics

(You can also find these in Shankar Chapter 4, where my "rules" are called "postulates" or Griffiths Chapter 3 "Generalized Statistical Interpretation", or in fact, in any Quantum text.)

Any quantum mechanical wave function must be normalizable, because the norm represents the total probability of finding the particle (or, more generally, the system) somewhere in its phase space, so

First Basic Rule: any state of the particle is a ket $|\psi\rangle$, symbolizing a function $\psi(x)$ in L2.

Mathematicians use the term Hilbert space to refer to inner-product spaces of normalizable functions such that any convergent sequence in the space has a limit in the space (a property that, for example, the rational numbers don't have, but the real numbers do) . Our functions above for the electron in the box do form such a space, with the sine waves an orthonormal basis. However, on going to the infinite line, although we still have normalizable wave functions, the two bases we have discussed above, the plane waves (momentum basis) and the delta functions (position basis) are not themselves in the space!

But these bases are both complete, meaning any wave function can be expressed in terms of a (continuous) sum over the elements of either of them.

Constructing these complete but not conventionally normalized bases was Dirac's doing, and is extremely convenient in describing quantum mechanics. But it upset the mathematicians. Fortunately, they later justified it by inventing the theory of distributions, which are generalized functions, and include delta functions.

Bottom Line: we shall follow the other physicists in using the term "Hilbert space" more loosely than mathematicians do, to refer to L2, including these non-normalizable bases.

Next Basic Rule: A physical variable, or observable, corresponds to a Hermitian operator A acting on L2.

We shall assume that the eigenkets of any such variable span the space: this is always true for a finite dimensional space, as previously discussed, but not for a general Hermitian operator in a Hilbert space, so this is a nontrivial assumption.

For an operator with a discrete set of eigenvalues, $A|n\rangle = \lambda_n|n\rangle$, any wave function can be written

$|\psi\rangle = \sum c_n|n\rangle$, with $c_n = \langle n|\psi\rangle$.

Rule for Relating Operators to Experiments: any measurement of the value of the physical variable A will yield one of the eigenvalues λ_n of the operator A, and the probability of finding the particular value λ_n is equal to $|c_n|^2 = |\langle n|\psi\rangle|^2$.

The expectation value of an observable A is the average value of a series of measurements on identical quantum systems,

$$\langle A\rangle = \langle\psi|A|\psi\rangle = \sum |c_n|^2\lambda_n.$$

It is important to note that two measurements of the same observable A on the same system, one measurement being made immediately after the other, must yield the same result. That is to say, if the first measurement reads λ_n, the second must be λ_n with 100 per cent probability. But this can only happen if the wave function after the first measurement is $|n\rangle$, which in general it wasn't before the first measurement. The jargon description of this is that the act of measurement "collapses the wave function" into one of the eigenstates of the variable being measured.

Measuring a Continuum Variable: For variables like position and momentum having continuum sets of eigenvectors, the statistical interpretation is in terms of finding the particle within some small range—the probability of finding it between x and x + dx is

$$\langle\psi|\int_x^{x+dx} dx|x\rangle\langle x|\psi\rangle = \int_x^{x+dx} |\psi(x)|^2 dx$$

and the expectation value of x is

$$\langle\psi|x|\psi\rangle = \langle\psi|\hat{x}\int_x^{x+dx} dx|x\rangle\langle x|\psi\rangle = \int_x^{x+dx} x|\psi(x)|^2 dx$$

where we've put a little hat on the x to remind us that it is an operator, with eigenkets.

Chapter 9

Schrodinger's Equation in 1-D

CURVATURE OF WAVE FUNCTIONS

Schrodinger's equation in the form

$$\frac{d^2\psi(x)}{dx^2} = \frac{2m(V(x)-E)}{\hbar^2}\psi(x)$$

can be interpreted by saying that the left-hand side, the rate of change of slope, is the curvature – so the curvature of the function is proportional to $(V(x)-E)\psi(x)$. This means that if E > V(x), for $\psi(x)$ positive $\psi(x)$ is curving negatively, for $\psi(x)$ negative $\psi(x)$ is curving positively. In both cases, $\psi(x)$ is always curving towards the x-axis—so, for E > V(x), $\psi(x)$ has a kind of stability: its curvature is always bringing it back towards the axis, and so generating oscillations. The simplest example is that of a constant potential V(x) = V0 < E, for which the wave function is $\psi(x) = A\sin(kx+\delta)$ with a constant and k = $\sqrt{(2m/\hbar^2)(E-V_0)}$.

On the other hand, for V(x) > E, the curvature is always away from the axis. This means that $\psi(x)$ tends to diverge to infinity. Only with exactly the right initial conditions will the curvature be just right to bring the wave function to zero as x goes to infinity. (This is possible because as $\psi(x)$ tends to zero, the curvature tends to zero, too.)

For a constant potential V0 > E, the wave function is $\psi(x) = Ae^{\alpha x} + Be^{-\alpha x}$, with $\alpha = \sqrt{(2m/\hbar^2)(V_0 - E)}$. Of course, this wave function will diverge in at least one direction! However, as we shall see below, there are situations with spatially varying potentials where this wave function is only relevant for positive x, and the coefficients A, B are functions of the energy—for certain energies it turns out that $A = 0$, and the wave function converges.

One Dimensional Infinite Depth Square Well

In an earlier lecture, we considered in some detail the allowed wave functions and energies for a particle trapped in an infinitely deep square well, that is, between infinitely high walls a distance L apart. For that case, the potential between the walls is identically zero so the wave function has the form $\psi(x) = A\sin(kx + \delta)$. The wave function $\psi(x)$ necessarily goes to zero right at the walls, since it cannot have a discontinuity, and must be zero just inside the wall. Even a quantum particle cannot penetrate an infinite wall!

An immediate consequence is that the lowest state cannot have zero energy, since k = 0 gives a constant $\psi(x)$. Rather, the lowest energy state must have the minimal amount of bending of the wave function necessary for it to be zero at both walls but nonzero in between—this corresponds to half a period of a sine or cosine (depending on the choice of origin), these functions being the solutions of Schrodinger's equation in the zero potential region between the walls. The allowed wave functions (eigenstates) found as the energy increases have successively 0, 1, 2, ... zeros (nodes) in the well.

Parity of a Wave Function

Notice that the allowed wave eigenfunctions of the Hamiltonian for the infinite well are symmetrical or antisymmetrical about the centre, $\psi(x) = \pm\psi(-x)$.

We call the operator that reflects a function in the origin the parity operator P, so these eigenstates of the Hamiltonian are also eigenstates of the parity operator, with eigenvalues 1. This is because the Hamiltonian is itself symmetric: d^2/dx^2 is even in x, and so is V(x), so $[H,P]=0$, and the two operators can be simultaneously diagonalized, that is, a common set of eigenstates can be constructed.

Finite Depth SquareWell

If the potential at the walls is not infinite, the parity operator P will continue to commute with the Hamiltonian H as long as the potential is symmetric, $V(x)=V(-x)$.

We take

V(x) = V0 for x < L/2

V(x) = 0 for L/2 < x < L/2

V(x) = V0 for L/2 < x.

We only need look for solutions symmetric or antisymmetric about the origin. This is important from a practical point of view, because it allows us to integrate Schrodinger's equation numerically out from the origin in the positive direction: $\psi(x)$ in the negative direction is fixed by symmetry (or antisymmetry). Since it's a second-order equation, we need two boundary conditions to get going, for symmetric states, we can take $\psi(x)=1,\ \psi'(x)=0$ at $x=0$, for antisymmetric states, $\psi(x)=0,\ \psi'(x)=1$ at $x=0$. (Of course, we will have to normalize $\psi(x)$ correctly eventually.)

The numerical strategy is to pick a value for the energy E, choose one of the boundary conditions above and integrate $\psi(x)$ numerically to a large positive value of x. For almost all values of E, the wave function will be exponentially increasing with x. For the particular values corresponding to bound states, it will be exponentially decreasing.

Let us begin by considering how the lowest energy state wave function is affected by having finite instead of infinite walls. Inside the well, where V = 0, the solution to Schrodinger's equation is still of cosine form (for a symmetric state). However, Schrodinger's equation now has a nonzero solution inside the wall $(x > L/2)$, where $V = V_0$:

$$-\frac{\hbar^2}{2m}\frac{d^2\psi(x)}{dx^2} + V_0\psi(x) = E\psi(x),$$

has two exponential solutions one increasing with x, the other decreasing,

$$e^{\alpha x} \text{ and } e^{-\alpha x}, \text{ where } \alpha = \sqrt{2m(V_0 - E)/\hbar^2}$$

We are assuming here that E < V0, so the particle is bound to the well. We shall find the lowest energy state is always bound in a finite square well, however weak the potential.

Now, Schrodinger's equation must be valid everywhere, including the point $x = L/2$. Since the potential is finite, the wave function $\psi(x)$ and its first derivative must be continuous at x = L/2.

Suppose, then, we choose a particular energy E. Then the wavefunction inside the well (taking the symmetric case) is proportional to coskx, where $k = \sqrt{2mE/\hbar^2}$. The wave function (and its derivative!) must match a sum of exponential terms at x = L/2, so

$$\cos(kL/2) = Ae^{\alpha L/2} + Be^{-\alpha L/2}$$

$$-k\sin(kL/2) = \alpha Ae^{\alpha L/2} - \alpha Be^{-\alpha L/2}$$

Solving these equations for the coefficients A, B in the usual way, we find that in general the cosine solution inside the well goes smoothly into a linear combination of exponentially increasing and decreasing terms in the wall. However, this cannot in general represent a bound state in the well. The increasing solution increases without limit as x goes to infinity, so since the square of the wave function is proportional to the probability of finding the particle at any point, the particle is infinitely more likely to be found at infinity than anywhere else. It got away! This clearly makes no sense – we're trying to find wave functions for particles that stay in, or at least close to, the well. We are forced to conclude that the only exponential wave function that makes sense is the one for which A is exactly zero, so that there is only a decreasing wave in the wall.

Finding the Bound State Energies Graphically

If we demand that the wavefunction decrease exponentially as x goes to infinity, or, in other words, require A to be zero, k must satisfy the condition given be dividing one of the boundary equations above by the other:

$$\tan(kL/2) = \alpha/k$$

This equation cannot be solved analytically, but is easy to solve graphically by plotting the two sides as functions of k (recall $\alpha = \sqrt{2m(V_0 - E)/\hbar^2}$, and $k = \sqrt{2mE/\hbar^2}$) and finding where the curves intersect.

From

$$\alpha^2 = \frac{2mV_0}{\hbar^2} - k^2 \quad \text{we find} \quad \frac{\alpha}{k} = \sqrt{\frac{2mV_0}{\hbar^2 k^2} - 1},$$

note that this is real only for

$$k < \frac{\sqrt{2mV_0}}{\hbar}.$$

(Because if this inequality is not satisfied, the particle has enough kinetic energy to get out of the well!)

Now the condition for a bound state can be written

$$\tan(kL/2) = \frac{\alpha}{k} = \sqrt{\frac{2mV_0}{\hbar^2 k^2} - 1}.$$

Cleaning up the appearance of the equation by choosing variables $x = \frac{kL}{2}$, $a^2 = \frac{mV_0L^2}{2\hbar^2}$, and plotting $\tan x$ and $\sqrt{\left(\frac{a}{x}\right)^2 - 1}$ from x = 0 to x = a, allowed bound state k-values correspond to the points of intersection of the two curves. The bound state energies are then given by $E = \hbar^2 k^2 / 2m$.

The variable a is a measure of the attractive strength of the well. Here are the two curves for a shallow well (a = 1):

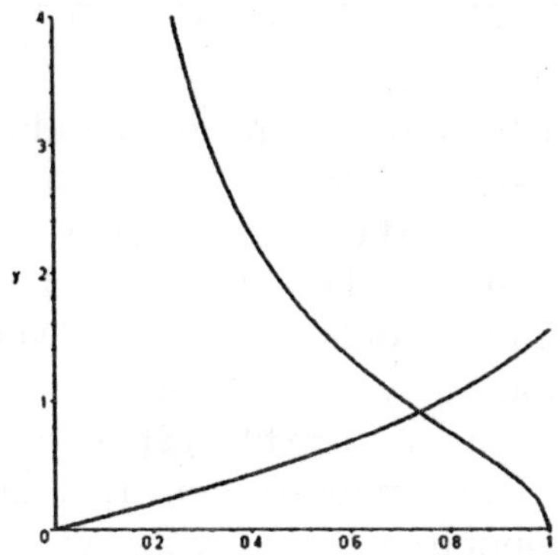

It is interesting to note that however small a is, the curve $y = \sqrt{\left(\frac{a}{x}\right)^2 - 1}$ goes to infinity as x goes to zero, so will always intersect y = tanx: there will always be a bound state.

A deeper well, a = 20, gives several bound states:

For the lower energies at least, the allowed k–values are approximately linearly spaced, at 1, 3, 5, ... so the bound state energies are not far off the 1, 9, 25,... pattern of the infinitely deep square well—remembering that we are only looking at the even parity (cosine) solutions!

The odd parity solutions, sine waves inside the well, can be found by an exactly similar analysis. One difference is that an arbitrarily weak well will not bind an odd parity state.

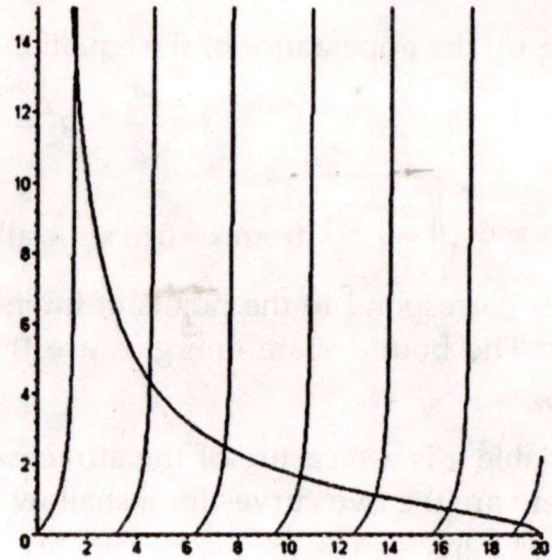

The point is that for a weak potential to bind an even state, it only has to curve the wave function slightly to get from one exponentially decaying to the left to one exponentially decaying to the right. These curves decay very slowly for a weak potential, and give a bound state in which the particle is most likely to be found outside the well. On the other hand, in an odd solution the wave function within the well has to have enough total curvature to fit together two decaying wave functions which have opposite sign. This takes much more bending, and cannot be achieved with a very weak potential.

THE DELTA FUNCTION POTENTIAL

One limiting case of a square well is a very narrow deep well, which can be approximated by a delta function when the range of variation of the wave function is much greater than the range of the potential, so Schrodinger's equation becomes

$$-\frac{\hbar^2}{2m}\frac{d^2\psi(x)}{dx^2}+\lambda\delta(x)\psi(x)=E\psi(x)$$

with λ negative for an attractive potential.

The infinity of the δ-function cannot be balanced by the finite right hand side, so the wave function must have a discontinuity in slope at the origin.

To find the ground state energy, note first that as a one-dimensional attractive potential there will be a bound state: any change in slope is sufficient to connect an exponentially increasing function coming in from $-\infty$ to a decreasing one going to $+\infty$, since the rates of increase and decrease can be arbitrarily slow.

Away from the origin, then, we can take the wave function to be

$$\psi(x) = Ae^{-\alpha|x|},$$

the energy of the state being $E = -\hbar^2\alpha^2/2m$.

The discontinuity in slope at the origin is just

$$\frac{d\psi(x=+\varepsilon)}{dx} - \frac{d\psi(x=-\varepsilon)}{dx} = -2A\alpha.$$

To match this with the δ-function singularity, we integrate the Schrodinger equation term by term from $-\varepsilon$ to $+\varepsilon$ in the limit of ε going to zero:

$$\int_{-\varepsilon}^{\varepsilon} -\frac{\hbar^2}{2m}\frac{d^2\psi(x)}{dx^2}dx + \int_{-\varepsilon}^{\varepsilon} \lambda\delta(x)\psi(x)dx = \int_{-\varepsilon}^{\varepsilon} E\psi(x)dx.$$

Note first that the right-hand side, having a finite integrand, must go to zero in the limit ε of going to zero.

The δ-function term must integrate to $\lambda\psi(0) = \lambda A$.

The first term just gives the discontinuity in slope,

$$\int_{-\varepsilon}^{\varepsilon} -\frac{\hbar^2}{2m}\frac{d^2\psi(x)}{dx^2}dx = -\frac{\hbar^2}{2m}\int_{-\varepsilon}^{\varepsilon}\frac{d}{dx}\left(\frac{d\psi}{dx}\right)dx = -\frac{\hbar^2}{2m}\left[\frac{d\psi}{dx}\right]_{-\varepsilon}^{\varepsilon} = \frac{\hbar^2}{2m}2A\alpha.$$

Schrodinger's equation is therefore satisfied if $\hbar^2\alpha/m + \lambda = 0$ (remembering λ is negative for an attractive potential).

The energy of the bound state is

$$E = -\frac{\hbar^2\alpha^2}{2m} = -\frac{m\lambda^2}{2\hbar^2}.$$

A Potential Step

Our analysis so far has been limited to real-valued solutions of the time-independent Schrodinger equation. This is fine for analyzing bound states in a potential, or standing waves in general, but cannot be used, for example, to represent an electron traveling through space after being emitted by an electron gun, such as in a TV tube. The reason is that a real-valued wave function $\psi(x)$, in an energetically allowed region, is made up of terms locally like coskx and sinkx, multiplied in the full wave function by the time dependent phase factor $e^{-iEt/\hbar}$, giving equal amplitudes of right moving waves $e^{i(kx-Et)/\hbar}$ and left moving waves $e^{-i(kx+Et)/\hbar}$. So, for an electron definitely moving to the right, even the time-independent part of the wave function must necessarily be complex.

Consider an electron of energy E moving in one dimension through a region of zero potential from large negative x and encountering an upward step potential of height V0 (V0 < E) at the origin x = 0. Of course, strictly speaking, the electron should be represented by a wave packet, and hence could not have a precisely defined energy E, but we assume here that it is a very long wave packet, very close to a plane wave, so we take it that the wave function is:

$$\psi(x,t) = Ae^{ikx}e^{-iEt/\hbar} \quad \text{for } x < 0$$

(A more precise analysis, in which an incoming wave packet is used, can be done by solving for the plane-wave components individually. In the limit of a wavepacket long compared to the de Broglie wavelength, the result is the same as that found here.)

Visualizing the classical picture of a particle approaching a hill (smoothing off the corners a bit) that it definitely has enough energy to surmount, we would perhaps expect that the wave function continues beyond x = 0 in the form

$$\psi(x,t) = Be^{ik_1x}e^{-iEt/\hbar} \quad \text{for } x > 0,$$

where k1 corresponds to the slower speed the particle will have after climbing the hill.

Schrodinger's equation requires that the wave function have no discontinuities and no kinks (discontinuities in slope) so the x < 0 and x > 0 wave functions must match smoothly at the origin. For them to have the same value, we see from above that A = B. For them to have the same slope we must have kA = k1B. Unfortunately, the only way to satisfy both these equations with our above wave functions is to take k = k1—which means there is no step potential at all!

we have been led astray by the depiction of the particles as little balls rolling along in a potential, with enough energy to get up the hill, etc. Schrodinger's equation is a wave equation. Building intuition about solutions should rely on experience with waves. We should be thinking about a light wave going from air into glass, for example. If we do, we realise that at any interface some of the light gets reflected. This means that our expression for the wave function for x < 0 is incomplete, we need to add a reflected wave, giving

$$\psi(x,t) = Ae^{ikx}e^{-iEt/\hbar} + Ce^{-ikx}e^{-iEt/\hbar} \quad \text{for } x < 0$$

$$\psi(x,t) = Be^{ik_1x}e^{-iEt/\hbar} \quad \text{for } x > 0.$$

Now matching the wave function and its derivative at the origin,

$$A + C = B$$

$$k(A - C) = k_1 B.$$

The fraction of the wave that is reflected

$$R = \frac{C^2}{A^2} = \left(\frac{k - k_1}{k + k_1}\right)^2.$$

Evidently, the fraction of the wave transmitted

$$T = 1 - R = \frac{4kk_1}{(k + k_1)^2}.$$

The ratio B2/A2 gives the relative probability of finding a particle in some small region in the transmitted stream relative to that in the incoming stream, but the particles in the transmitted stream are moving more slowly, by a factor k1/k. That means that just comparing the densities of particles in the transmitted and incoming streams is not enough. The physically significant quantity is the probability current flowing past a given point, and this is the product of the density and the speed. Therefore, the transmission coefficient is B2k1/A2k.

Tunneling through a Square Barrier

If a plane wave coming in from the left encounters a step at the origin of height V0 > E, the incoming energy, there will be total reflection, but with an exponentially decaying wave penetrating some distance into the step. Suppose now we replace the step with a barrier,

V = 0 for x < 0, call this region I

V = V0 for 0 < x < L, this is region II

V = 0 for L < x, region III.

In this situation, the wave function will still decay exponentially into the barrier (assuming the barrier is thick compared to the exponential decay length), but on reaching the far end at x = L, a plane wave solution is again allowed, so there is a nonzero probability of finding the particle beyond the barrier, moving with its original speed. This phenomenon is called tunneling, since in the classical picture the particle doesn't have enough energy to get over the top of the barrier.

The way to solve the problem is to solve the Schrodinger equation in the three regions, then apply the boundary conditions. Since we are interested in the probability of a particle getting through the barrier, we do not need to worry about normalizing the wave function, so for simplicity we take an incoming wave of unit amplitude. In region I, there will also be a reflected wave, so

$$\psi_I(x) = e^{ikx} + Be^{-ikx}.$$

In region II, there will in general be both exponentially decreasing and exponentially growing solutions, so we take

$$\psi_{II}(x) = Ce^{\alpha x} + De^{-\alpha x}$$

Recall $\alpha = \sqrt{(2m/\hbar^2)(V_0 - E)}$.

In region III, there is only the outgoing wave, to make the equations easy we absorb a phase factor in the coefficient, and write:

$$\psi_{III}(x) = Se^{ik(x-L)}.$$

We now require and be continuous at x = 0, L. Elementary computations lead to

$$2ik = (ik+\alpha)C + (ik-\alpha)D$$

$$2\alpha Ce^{\alpha L} = (ik+\alpha)S$$

$$2\alpha De^{-\alpha L} = -(ik-\alpha)S.$$

Solving these equations gives

$$S(k) = \frac{2k\alpha}{2k\alpha\cosh\alpha L + i(\alpha^2 - k^2)\sinh\alpha L.}$$

The probability of tunneling is | S |2,

$$|S(k)|^2 = \frac{4k^2\alpha^2}{(k^2+\alpha^2)^2\sinh^2\alpha L + 4k^2\alpha^2}.$$

An important limit is that of a barrier thick compared with the decay length, $\alpha L \gg 1$.

In this limit, $\sinh^2\alpha L \approx \frac{1}{4}e^{-2\alpha L}$, and using $\alpha = \sqrt{(2m/\hbar^2)(V_0 - E)}$, $E = \hbar^2 k^2 / 2m$, we find

$$|S(k)|^2 \cong 16\left(\frac{E}{V_0}\right)\left(1 - \frac{E}{V_0}\right)e^{-2\alpha L}$$

In typical tunneling problems, the far and away dominant term is the $e^{-2\alpha L}$, which may differ from unity by many orders of magnitude.

The Spherically Symmetric Three-Dimensional Problem

The methods developed above for the one-dimensional system are almost immediately applicable to a very important three-dimensional case: a particle in a spherically symmetric potential.We restrict ourselves here to spherically symmetric solutions of Schrodinger's equation $\psi(r,\theta,\phi)=\psi(r)$, a subspace of the space of all possible solutions that always includes the ground state.

The kinetic energy operator on states in this subspace (where $d/d\theta \equiv 0$, $d/d\phi \equiv 0$) is

$$\frac{\vec{p}^2}{2m}=-\frac{\hbar^2\vec{\nabla}^2}{2m}=-\frac{\hbar^2}{2m}\frac{1}{r^2}\frac{d}{dr}\left(r^2\frac{d}{dr}\right).$$

It is easy to check that if we write the wave function

$$\psi(r)=\frac{u(r)}{r}$$

the function u(r) obeys the one-dimensional equation

$$-\frac{\hbar^2}{2m}\frac{d^2u(r)}{dr^2}+V(r)u(r)=Eu(r)$$

exactly like a particle in one dimension, except that here r is only positive, and u(r) must go to zero at the origin.

Exercise: for a spherical square well, $V=0$ if $r<r_0$, $V=V_0$ for $r\geq r_0$, find the minimum value of V0 for which a bound state exists for given r0 and particle mass m. Sketch the wave function.

Alpha Decay

A good example of tunneling, and one which historically helped establish the validity of quantum ideas at the nuclear level, is α-decay. Certain large unstable nuclei decay radioactively by emitting an α-particle, a tightly bound state of two protons and two neutrons. It is thought that α-particles may exist, at least as long lived resonances, inside the nucleus. For such a particle, the strong but short ranged nuclear force creates a spherical finite depth well having a steep wall more or less coinciding with the surface of the nucleus. However, we must also include the electrostatic repulsion between the α-particle and the rest of the nucleus, a potential $(Z-2)2e^2/r$ outside the nucleus. This means that, as seen from inside the nucleus, the wall at the surface may not be a step but a barrier, in the sense we used the word above, a step up followed by a slide down the electrostatic curve:

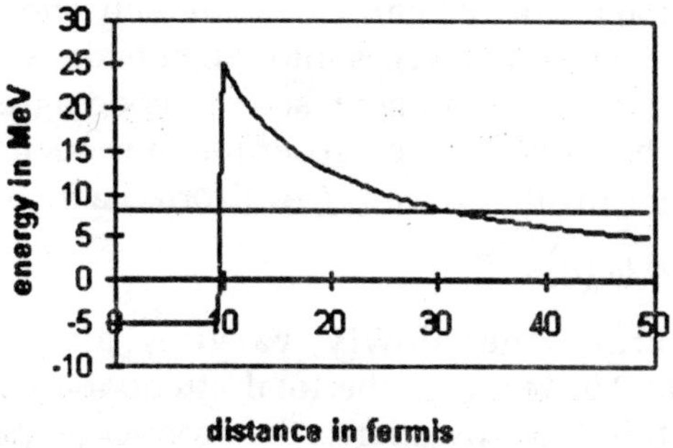

Therefore, an α-particle bouncing around inside the nucleus may have enough energy to tunnel through to the outside world.

The α-particles are emitted with spherical symmetry, so the wave function can be written $\psi(r)=u(r)/r$, as discussed above, and Schrodinger's equation is

$$-\frac{\hbar^2}{2m}\frac{d^2u(r)}{dr^2}+V(r)u(r)=Eu(r), \quad r>0.$$

It is evident that the more energetic the α-particle is, the thinner the barrier it faces. Since the wave function decays exponentially in the barrier, this can make a huge difference in tunneling rates. It is not difficult to find the energy with which the α-particle hits the nuclear wall, because this will be the same energy with which it escapes. Therefore, if we measure the energy of an emitted α, since we think we know the shape of the barrier pretty well, we should be able, at least numerically, to predict the tunneling rate. The only other thing we need to know is how many times per second α's bounce off the wall. The size of the nucleus is of order 10-14 meters (10 fermis), if we assume an α moves at, say, 107 meters per second, it will bang into the wall 1021 times per second. This is a bit handwaving, but all α-radioactive nuclei are pretty much the same size, so perhaps it's safe to assume this will be about the same for all of them.

This is a huge number – the probability of transmission is evidently very tiny! In other words, the decay length of the wave function inside the barrier is extremely short (except for the very last bit as it emerges into the outside world). It's so short, in fact, that we can get results in good agreement with experiment by dividing the barrier into a sequence of square barriers and using the above $\alpha L \gg 1$ formula for each of them,

$$\psi(r+\Delta r) \approx \psi(r) e^{-\alpha(r)\Delta r}$$

(neglecting the slowly varying and completely unimportant 1/r term) so the total attenuation of the wave function in this approximation is the product of terms like this,

$$\psi(\text{exit}) = \psi(\text{nucl}) \prod_i e^{-\alpha(r_i)\Delta r}$$

$$\approx \psi(\text{nucl}) e^{-\int \alpha(r) dr}$$

$$= \psi(\text{nucl}) e^{-\int \sqrt{(2m/\hbar^2)(V(r)-E)} dr}$$

Since we know the actual shape of the barrier, this integral can be done with elementary functions, and the result for the transmission rate is

$$|S(k)|^2 \approx Ae^{-B/\sqrt{E}}.$$

Here $B = \left(\pi\sqrt{2m}/\hbar\right)(Z-2)2e^2$ is not an adjustable parameter: and plotting $\ln|S(k)|^2$ against $1/\sqrt{E}$ for

Polonium212 (which emits α's with energy 8.95MeV, and lasts 3×10^{-7} seconds) Thorium232 (4.05MeV α's, 1.4×10^{10} years), and several intermediate lifetime nuclei gives a straight line plot with the correct slope within a few percent! These elements can all be understood in terms of essentially the same barrier being tunneled through at the different heights corresponding to the α energy.Further refinements make little difference to the final result in this case.

Chapter 10

The Energy-Time Uncertainty Principle

MODEL OF A DECAYING STATE

The momentum-position uncertainty principle $\Delta p \cdot \Delta x \geq \hbar$ has an energy-time analog, $\Delta E \cdot \Delta t \geq \hbar$. Evidently, though, this must be a different kind of relationship to the momentum-position one, because t is not a dynamical variable, so this can't have anything to do with non-commutation.

To illustrate the meaning of the equation $\Delta E \cdot \Delta t \geq \hbar$, let us reconsider α-decay, but with a slightly simplified potential to clarify what's going on. Specifically, we replace the combined nuclear force/electrostatic repulsion barrier with a square barrier, high enough and thick enough that there is a small probability per unit time of the particle tunneling out of the well.

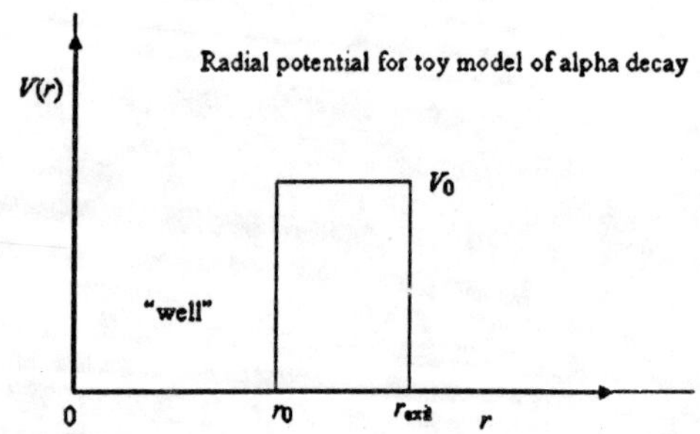

If the barrier thickness were increased to infinity (keeping r0 fixed) there would be a true bound state having energy E0, and for E0 well below V0, having approximately an integral number of half wavelengths in the well. For a barrier of finite thickness, there is of course some nonzero probability of the particle escaping—so no longer a true bound state, but for a thick barrier the difference may be hard to detect.

As with the α-decay analysis, we'll look at this semiclassically, picturing the particle as bouncing off the walls backwards and forwards inside, time τ between hits, and at each hit probability of penetration some small number ε. Therefore, the probability that the particle is still in the well after a time $t = n\tau$ is $P(n\tau) = (1-\varepsilon)^n$. Since ε really is very small for α-decay (less than 10-12), we can conveniently write P as a function of time by using the formula $e^{-x} = \lim_{\varepsilon\to 0}(1-\varepsilon)^{x/\varepsilon}$.

From this, the probability of the particle being in the well after n bounces, time $t = n\tau$,

$$P(n\tau) = (1-\varepsilon)^n = (1-\varepsilon)^{\varepsilon n/\varepsilon} \cong e^{-\varepsilon n}.$$

The standard notation is:

$$P(t) = e^{-\Gamma t/\hbar}$$

where $\Gamma t/\hbar = \Gamma n\tau/\hbar = \varepsilon n$.

Note that $\Gamma = \hbar\varepsilon/\tau$ has the dimensions of energy (ε being dimensionless). If $\varepsilon/\tau \ll 1$, is the decay rate—the probability of decay per second; Γ is essentially an inverse lifetime.

The exponential decay law for radioactive elements is completely confirmed experimentally, it is the basis of the "half-life" rule: for any given amount of a radioactive nucleus, half of it will decay in a time period the half-life—fixed for that nucleus.

Obviously, if the modulus of the wave function of the particle in the well is decreasing with time, the time dependence of ψ can no longer be just the $e^{-iE_0t/\hbar}$ of the original "bound state". From P(t) above, evidently we must have

$$|\psi(t)|^2 = |\psi(0)|^2 e^{-\Gamma t/\hbar}.$$

This is a far slower time dependence than that of the $e^{-iE_0t/\hbar}$ term, so it is an excellent approximation to put the time dependences together in one exponential factor:

$$\psi(t) = \psi(0)e^{-i(E_0-i\Gamma/2)t/\hbar}$$

At this point, the analogy with $\Delta p \cdot \Delta x \geq \hbar$ emerges. Recall we introduced the p, x uncertainty principle by finding what spread Δp in the Fourier components of a wave packet were necessary in order for the wave packet to die away from its centre over a distance of order Δx. A true localized wave packet has a continuum of p components, but the right expression for the spread in momentum space Δp turns out to be given by taking just two waves, $e^{ipx/\hbar}$ and $e^{i(p+\Delta p)x/\hbar}$, and noticing that they fall out of phase in a distance $\Delta x \sim \hbar/\Delta p$. The more precise derivation based on Gaussian wave packets reaches essentially the same conclusion.

Now, in the present situation the wave function decays in time rather than space, but the argument is very similar. To construct the decaying wave function we must add together "plane waves in time" $e^{-iEt/\hbar}$ corresponding to different energies. The required spread in energy ΔE follows from an argument just like the one above for space: if the wave function is to be die away in a time of order Δt, in other words to be "localized in time", it must be constructed of waves having a range ΔE in energies such that $e^{-iEt/\hbar}$ and $e^{-i(E+\Delta E)t/\hbar}$ get out of sync in a time Δt. This gives immediately $\Delta E \cdot \Delta t \geq \hbar$.

It's worth looking a little further into just what superposition of energy "plane waves" $e^{-iEt/\hbar}$ gives the required exponential-in-time behaviour of the wave function for α-decay. Writing

$$\psi(t)=\psi(0)\int c(E)e^{-iEt/\hbar}dE/2\pi$$

the Fourier coefficients are given by

$$c(E)=\int_0^{\infty} e^{i(E-E_0+i\Gamma/2)t/\hbar}dt=\frac{i\hbar}{E-E_0+i\Gamma/2}.$$

This distribution (called Lorentzian) has a narrow peak of width of order Γ and height of order $1/\Gamma$ centered at E0. (Strictly speaking, this is an approximation in that c(E) must of course be zero for E negative—we ignore that tiny correction here.)

In fact, for the α-decaying nucleus, this energy spread is undetectably small, but that is certainly not the case for other decaying states, where this same analysis applies. In particular, some of the resonant states created in collisions of elementary particles have masses of order 1,000 Mev, and lifetimes of order 10-23 seconds—corresponding to a width Γ of the order of 10 per cent of E0! Obviously, these transient bound states are far from eigenstates of the Hamiltonian—but you will find them listed in particle tables.

RESONANCES

The time-reversed wave function for α-decay is also a perfectly good solution to Schrodinger's equation. In principle, if we could arrange for an α-particle to have a spherical ingoing wave function within the narrow energy range corresponding to the quasi bound state, ψ would become very large inside the nucleus, meaning that the α-particle would spend a very long time there compared with the time spent at any other point on the way in: this is called resonance. This particular experiment is never likely to take place, but precisely analogous experiments involving particles scattering off each other, and electrons scattering resonantly from atoms, are common.

One can interpret the wave function for a true bound state as a standing wave, radially containing a whole number of half wavelengths, so that when the wave is reflected at the walls it has just the right phase to interfere constructively with itself. A resonance will have the same wavelength requirement (but the reflection at the walls is of course no longer perfect).

We shall see that in the case of nonzero angular momentum, the centrifugal barrier provides an effective repulsive term, which together with an attractive force can create a barrier configuration having similar effect to that in the toy model.

The pictures below are for the energy right at resonance for two different barrier heights. Evidently, if the height of the barrier is increased, $|\psi|^2$ inside at resonance increases as well. This is easy to understand—the higher barrier is more difficult to penetrate, so the particle spends even longer inside. The lifetime Δt is clearly proportional to $|\psi|^2$ inside.

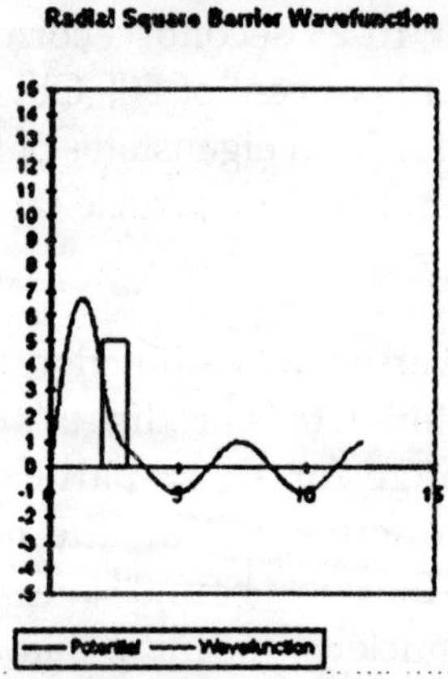

This energy spread of the resonance is the Γ above, and is related to the "lifetime" of the resonance by $\Delta E \cdot \Delta t \geq \hbar$. How does the ratio of the two Γ's for these two barriers relate to the ratio of the two maximum values of $|\psi|^2$? Give a physical explanation of your findings.

Chapter 11

General Uncertainty Principle

UNCERTAINTY AND NON-COMMUTATION

As we discussed in the Linear Algebra lecture, if two physical variables correspond to *commuting* Hermitian operators, they can be diagonalized simultaneously—that is, they have a common set of eigenstates. In these eigenstates *both* variables have precise values at the same time, there is no "Uncertainty Principle" requiring that as we know one of them more accurately, we increasingly lose track of the other. For example, the energy and momentum of a free particle can both be specified exactly. More interesting examples will appear in the sections on angular momentum and spin.

But if two operators do *not* commute, in general one *cannot* specify both values precisely. Of course, such operators could still have *some* common eigenvectors, but the interesting case arises in attempting to measure A and B simultaneously for a state $|\psi\rangle$ in which the commutator $[A,B]$ has a nonzero expectation value,

$$\langle\psi|[A,B]|\psi\rangle \neq 0$$

A QUANTITATIVE MEASURE OF "UNCERTAINTY"

Our task here is to give a *quantitative* analysis of how accurately noncommuting variables can be measured together. We found earlier using a semi-quantitative argument that for a free particle, $\Delta p \cdot \Delta x \sim \hbar$ at best. To improve on that result, we need to be precise about the uncertainty ΔA in a state $|\psi\rangle$.

We *define* ΔA as the root mean square deviation:

$$(\Delta A)^2 = \langle\psi|(A-\langle A\rangle)^2|\psi\rangle, \text{ where } \langle A\rangle = \langle\psi|A|\psi\rangle.$$

To make the equations more compact, we define $\hat{a}$ by

$$A = \langle A\rangle + \hat{a}.$$

(We'll put a caret (a hat) on the $\hat{a}$ to remind ourselves it's an operator—and, of course, it's a *Hermitian* operator, like *A*.) We also drop the ψ bra and ket, on the understanding that this whole argument is for a particular state. Now

$$(\Delta A)^2 = \left\langle (A-\langle A\rangle)^2 \right\rangle = \left\langle \hat{a}^2 \right\rangle.$$

Introduce an operator B in exactly similar fashion, $B = \langle B\rangle + \hat{b}$, having the property that $\langle\psi|[A,B]|\psi\rangle \neq 0$.

THE GENERALIZED UNCERTAINTY PRINCIPLE

The quantitative measure of how the combined "uncertainty" of measuring two variables relates to their lack of commutativity is most simply presented as a

THEOREM

$$(\Delta A)^2(\Delta B)^2 \geq \tfrac{1}{4}\langle i[A,B]\rangle^2.$$

(*Remark*: remember that for *A*, *B* Hermitian, $[A,B]$ is *anti*Hermitian: so is real! To make this clear, take adjoints: if *H* is hermitian then $(\langle\psi|H|\psi\rangle)^* = \langle\psi|H^\dagger|\psi\rangle = \langle\psi|H|\psi\rangle$, so $\langle\psi|H|\psi\rangle$ is real. If *K* is *anti* Hermitian, $K^\dagger = -K$, then $(\langle\psi|K|\psi\rangle)^* = \langle\psi|K^\dagger|\psi\rangle = -\langle\psi|K|\psi\rangle$, from which $\langle\psi|K|\psi\rangle$ is *pure imaginary*.)

PROOF OF THE THEOREM:

Define

$$|\psi_a\rangle = \hat{a}|\psi\rangle, \quad |\psi_b\rangle = \hat{b}|\psi\rangle.$$

Then

$$(\Delta A)^2(\Delta B)^2 = \langle\psi|\hat{a}^2|\psi\rangle\langle\psi|\hat{b}^2|\psi\rangle = \langle\psi_a|\psi_a\rangle\langle\psi_b|\psi_b\rangle$$

Using Schwartz's inequality

$$\langle\psi_a|\psi_a\rangle\langle\psi_b|\psi_b\rangle \geq |\langle\psi_a|\psi_b\rangle|^2$$

gives immediately

$$(\Delta A)^2(\Delta B)^2 \geq |\langle\psi_a|\psi_b\rangle|^2 = |\langle\psi|\hat{a}\hat{b}|\psi\rangle|^2.$$

The operator $\hat{a}\hat{b}$ is neither Hermitian nor antiHermitian. To evaluate the mod squared of its expectation value, we break the amplitude into real and imaginary parts:

$$\langle\psi|\hat{a}\hat{b}|\psi\rangle = \langle\psi|\tfrac{1}{2}\left(\hat{a}\hat{b}+\hat{b}\hat{a}\right)|\psi\rangle + \langle\psi|\tfrac{1}{2}\left[\hat{a},\hat{b}\right]|\psi\rangle.$$

(The first term on the right-hand side is the expectation value of a Hermitian matrix, and so is real, the second term is the expectation value of an antiHermitian matrix, so is pure imaginary.)

It follows immediately that

$$|\langle\psi|\hat{a}\hat{b}|\psi\rangle|^2 \geq |\langle\psi|\tfrac{1}{2}\left[\hat{a},\hat{b}\right]|\psi\rangle|^2.$$

But $[A,B] = [\hat{a},\hat{b}]$, so

$$(\Delta A)^2(\Delta B)^2 \geq \tfrac{1}{4}\langle i[A,B]\rangle^2.$$

MINIMIZING THE UNCERTAINTY

For a particle in one dimension denote

$$A = x, \quad B = p = -i\hbar\frac{d}{dx}, \text{ so } [A,B] = -i\hbar\left(x\frac{d}{dx} - \frac{d}{dx}x\right) = i\hbar.$$

(It important in that last step to understand that the operator $\frac{d}{dx}$ operates on everything to its right, and, as we are always finding matrix elements of operators, there will be a following ket it operates on, so $\frac{d}{dx}x = 1 + x\frac{d}{dx}$.)

We conclude that

$$(\Delta x)^2 (\Delta p)^2 \geq \tfrac{1}{4}\hbar^2.$$

Is there a wavefunction for which this inequality becomes an equality?

That would require $|\langle\psi_a|\psi_b\rangle|^2 = \langle\psi_a|\psi_a\rangle\langle\psi_b|\psi_b\rangle$, which can only be true if the two vectors are parallel, $|\psi_b\rangle = \lambda|\psi_a\rangle$, or, written explicitly,

$$\left(-i\hbar\frac{d}{dx} - \langle p\rangle\right)\psi(x) = \lambda\left(x - \langle x\rangle\right)\psi(x).$$

Actually , that's not enough: we *also* need $\langle\psi|\tfrac{1}{2}\left(\hat{a}\hat{b}+\hat{b}\hat{a}\right)|\psi\rangle$ to be zero. (Look at the equation above giving $\langle\psi|\hat{a}\hat{b}|\psi\rangle$ in terms of its real and imaginary parts, and how we used it to establish the inequality.)

Writing $|\psi_b\rangle = \lambda|\psi_a\rangle$ as $\hat{b}|\psi\rangle = \lambda\hat{a}|\psi\rangle$ and $\langle\psi|\hat{b} = \lambda^*\langle\psi|\hat{a}$ we find

$$\langle\psi|\tfrac{1}{2}\left(\hat{a}\hat{b}+\hat{b}\hat{a}\right)|\psi\rangle = \left(\lambda+\lambda^*\right)\langle\psi|\hat{a}^2|\psi\rangle,$$

so this will be zero if and only if λ is pure imaginary.

Turning to the differential equation, we first take the simplest case where $\langle x\rangle$ and $\langle p\rangle$ are both zero. The first requirement just sets the origin, but the second stipulates that our wave function has no net momentum.

For this simple case, $|\psi_b\rangle = \lambda|\psi_a\rangle$ becomes

$$-i\hbar\frac{d\psi(x)}{dx} = \lambda x\psi(x)$$

$$\frac{d\psi}{\psi} = \frac{i\lambda}{\hbar}xdx$$

$$\psi = Ce^{i\lambda x^2/2\hbar}$$

and recalling that λ is pure imaginary, this is a Gaussian wave packet! It is straightforward to check that the solution with $\langle x\rangle$ and $\langle p\rangle$ nonzero is

$$\psi(x) = Ce^{i\langle p\rangle x/\hbar}e^{-\alpha(x-\langle x\rangle)^2/2\hbar}$$

where $\alpha = -i\lambda$ is real, and C is the usual Gaussian normalization constant.

The conclusion is then that the Gaussian wave packet gives the optimum case for minimizing the joint uncertainties in position and momentum.

Note that the condition $\hat{b}|\psi\rangle = \lambda\hat{a}|\psi\rangle$ does *not* mean that $|\psi\rangle$ is an eigenstate of either $\hat{a}$ or $\hat{b}$, but it *is* an eigenstate of the *nonHermitian* operator , with eigenvalue zero. We shall soon see that this nonHermitian operator $\hat{b} - \lambda\hat{a} = \hat{b} + i\alpha\hat{x}$ and its adjoint play important roles in the quantum mechanics of the simple harmonic oscillator.

Chapter 12

The Simple Harmonic Oscillator

EINSTEIN'S SOLUTION OF THE SPECIFIC HEAT PUZZLE

The simple harmonic oscillator, a nonrelativistic particle in a potential $\frac{1}{2}kx^2$, is an excellent model for a wide range of systems in nature. In fact, not long after Planck's discovery that the black body radiation spectrum could be explained by assuming energy to be exchanged in quanta, Einstein applied the same principle to the simple harmonic oscillator, thereby solving a long-standing puzzle in solid state physics—the mysterious *drop in specific heat* of all solids at low temperatures. Classical thermodynamics, a very successful theory in many ways, predicted no such drop—with the standard equipartition of energy, kT in each mode (potential plus kinetic), the specific heat should remain more or less constant as the temperature was lowered (assuming no phase change).

To explain the anomalous low temperature behaviour, Einstein assumed each atom to be an independent (quantum) simple harmonic oscillator, and, just as for black body radiation, he assumed the oscillators could only absorb or emit energy in *quanta*. Consequently, at low enough temperatures there is rarely sufficient energy in the ambient thermal excitations to excite the oscillators, and they freeze out, just as blue oscillators do in low temperature black body radiation. Einstein's picture was later somewhat refined—the basic set of oscillators was taken to be standing sound wave oscillations in the solid rather than individual atoms (making the picture even more like black body radiation in a cavity) but the main conclusion—the drop off in specific heat at low temperatures—was not affected.

THE CLASSICAL SIMPLE HARMONIC OSCILLATOR

The classical equation of motion for a one-dimensional simple harmonic oscillator with a particle of mass *m* attached to a spring having spring constant *k* is

$$m\frac{d^2x}{dt^2} = -kx.$$

The solution is

$$x = x_0 \sin(\omega t + \delta), \quad \omega = \sqrt{\frac{k}{m}},$$

and the momentum $p = mv$ has time dependence

$$p = mx_0\omega \cos(\omega t + \delta).$$

The total energy

$$(1/2m)(p^2 + m^2\omega^2x^2) = E$$

is clearly constant in time.

It is often useful to picture the time-development of a system in *phase space*, in this case a two-dimensional plot with position on the *x*-axis, momentum on the *y*-axis. Actually, to have (x,y)coordinates with the same dimensions, we use $(m\omega x, p)$.

It is evident from the above expression for the total energy that in these variables the point representing the system in phase space moves clockwise around a circle of radius $\sqrt{2mE}$ centered at the origin.

Note that in the *classical* problem we could choose any point $(m\omega x, p)$, place the system there and it would then move in a circle about the origin. In the *quantum* problem, on the other hand, we cannot specify the initial coordinates $(m\omega x, p)$ precisely, because of the uncertainly principle. The best we can do is to place the system initially in a small cell in phase space, of size $\Delta x \cdot \Delta p = \hbar / 2$. In fact, we shall find that in quantum mechanics phase space is always divided into cells of essentially this size for each pair of variables.

SCHRODINGER'S EQUATION AND THE GROUND STATE WAVE FUNCTION

From the classical expression for total energy given above, the Schrodinger equation for the quantum oscillator follows in standard fashion:

$$-\frac{\hbar^2}{2m}\frac{d^2\psi(x)}{dx^2}+\frac{1}{2}m\omega^2x^2\psi(x)=E\psi(x)$$

What will the solutions to this Schrodinger equation look like? Since the potential $\frac{1}{2}m\omega^2x^2$ increases without limit on going away from $x = 0$, it follows that no matter how much kinetic energy the particle has, for sufficiently large x the potential energy dominates, and the (bound state) wavefunction decays with increasing rapidity for further increase in x. (Obviously, for a real physical oscillator there is a limit on the height of the potential—we will assume that limit is much greater than the energies of interest in our problem.)

We know that when a particle penetrates a barrier of constant height V_0 (greater than the particle's kinetic energy) the wave function decreases exponentially into the barrier, as $e^{-\alpha x}$, where $\alpha = \sqrt{2m(V_0 - E)/\hbar^2}$. But, in contrast to this constant height barrier, the "height" of the simple harmonic oscillator potential *continues to increase* as the particle penetrates to larger x. Obviously, in this situation the decay will be faster than exponential. If we (rather naïvely) assume it is more or less *locally* exponential, but with a local α varying with V_0, neglecting E relative to V_0 in the expression for

α suggests that α itself is proportional to x (since the potential is proportional to x^2, and $\alpha \propto \sqrt{V}$) so maybe the wavefunction decays as $e^{-(\text{constant})x^2}$?

To check this idea, we insert $\psi(x) = e^{-x^2/2b^2}$ in the Schrodinger equation, using

$$\frac{d^2\psi}{dx^2} = -\frac{1}{b^2}\psi + \frac{x^2}{b^4}\psi$$

to find

$$-\frac{\hbar^2}{2m}\left(-\frac{1}{b^2} + \frac{x^2}{b^4}\right)\psi(x) + \frac{1}{2}m\omega^2x^2\psi(x) = E\psi(x).$$

The $\psi(x)$ is just a factor here, and it is never zero, so can be cancelled out. This leaves a quadratic expression which must have the same coefficients of x^0, x^2 on the two sides, that is, the coefficient of x^2 on the left hand side must be zero:

$$\frac{\hbar^2}{2mb^4} = \frac{m\omega^2}{2}, \text{ so } b = \sqrt{\frac{\hbar}{m\omega}}.$$

This fixes the wave function. Equating the constant terms fixes the energy:

$$E = \frac{\hbar^2}{2mb^2} = \frac{1}{2}\hbar\omega.$$

So the conjectured form for the wave function is in fact the *exact* solution for the lowest energy state! (It's the lowest state because it has no nodes.)

Also note that even in this ground state the energy is *nonzero*, just as it was for the square well. The central part of the wave function must have some curvature to join together the decreasing wave function on the left to that on the right. This "zero point energy" is sufficient in one physical case to melt the lattice—helium is liquid even down to absolute zero temperature (checked down to microkelvins!) because the wave function spread destabilizes the solid lattice that will form with sufficient external pressure.

HIGHER ENERGY STATES

It is clear from the above discussion of the ground state that $b = \sqrt{\frac{\hbar}{m\omega}}$ is the natural unit of length in this problem, and $\hbar\omega$ that of energy, so to investigate higher energy states we reformulate in dimensionless variables,

$$\xi = \frac{x}{b} = x\sqrt{\frac{m\omega}{\hbar}}, \quad \varepsilon = \frac{E}{\hbar\omega}.$$

Schrodinger's equation becomes

$$\frac{d^2\psi(\xi)}{d\xi^2} = (\xi^2 - 2\varepsilon)\psi(\xi).$$

Deep in the barrier, the ε term will become negligible, and just as for the ground state wave function, higher bound state wave functions will have $e^{-\xi^2/2}$ behaviour, multiplied by some more slowly varying factor (it turns out to be a polynomial).

Exercise: find the relative contributions to the second derivative from the two terms in $x^n e^{-x^2/2}$. For given *n*, when do the contributions involving the first term become small? Define "small". The standard approach to solving the general problem is to factor out the $e^{-\xi^2/2}$ term, $\psi(\xi) = h(\xi)e^{-\xi^2/2}$ giving a differential equation for $h(\xi)$: $\frac{d^2h}{d\xi^2} - 2\xi\frac{dh}{d\xi} + (2\varepsilon - 1)h = 0$ We try solving this with a power series in ξ: $h(\xi) = h_0 + h_1\xi + h_2\xi^2 = \ldots$.

Inserting this in the differential equation, and requiring that the coefficient of each power ξ^n vanish identically, leads to a recurrence formula for the coefficients h_n:

$$h_{n+2} = \frac{(2n+1-2\varepsilon)}{(n+1)(n+2)} h_n .$$

Evidently, the series of odd powers and that of even powers are independent solutions to Schrodinger's equation. (Actually this isn't surprising: the potential is even in x, so the parity operator P commutes with the Hamiltonian. Therefore, unless states are degenerate in energy, the wave functions will be even or odd in x.) For large n, the recurrence relation simplifies to

$$h_{n+2} \approx \frac{2}{n} h_n, \quad n \gg \varepsilon.$$

The series therefore tends to

$$\sum \frac{2^n \xi^{2n}}{(2n-2)(2n-4)\dots 2} = 2\xi^2 \sum \frac{\xi^{2(n-1)}}{(n-1)!} = e^{\xi^2}.$$

Multiply this by the $e^{-\xi^2/2}$ factor to recover the full wavefunction, we find ψ diverges for large x as $e^{+\xi^2/2}$.

Actually we should have expected this—for a general value of the energy, the Schrodinger equation has the solution $\approx Ae^{+\xi^2/2} + Be^{-\xi^2/2}$ at large distances, and only at certain energies does the coefficient A vanish to give a normalizable bound state wavefunction.

So how do we find the *nondiverging* solutions? It is clear that the infinite power series must be stopped! The key is in the recurrence relation. If the energy satisfies

$$2\varepsilon = 2n+1, \quad n \text{ an integer},$$

then h_{n+2} and all higher coefficients vanish. This requirement in fact completely determines the polynomial (except for an overall constant) because with $2\varepsilon = 2n+1$ the coefficients h_m for $m < n$ are determined by

$$h_{m+2} = \frac{(2m+1-2\varepsilon)}{(m+1)(m+2)} h_m = \frac{(2m+1-(2n+1))}{(m+1)(m+2)} h_m .$$

This n^{th} order polynomial is called a *Hermite polynomial* and written $H_n(\xi)$. The standard normalization of the Hermite polynomials $H_n(\xi)$ is to take the coefficient of the highest power ξ^n to be 2^n. The other coefficients then follow using the recurrence relation above, giving:

$H_0(\xi)=1,\ H_1(\xi)=2\xi,\ H_2(\xi)=4\xi^2-2,\ H_3(\xi)=8\xi^3-12\xi,$ etc.

So the bottom line is that the wavefunction for the nth excited state, having energy $\varepsilon = n+\frac{1}{2}$, is $\psi_n(\xi)=C_nH_n(\xi)e^{-\xi^2/2}$, where C_n is a normalization constant to be determined in the next section.It can be shown (see exercises at the end of this lecture) that $H_n'(\xi)=2nH_{n-1}(\xi)$. Using this, beginning with the ground state, one can easily convince oneself that the successive energy eigenstates each have one more node—the nth state has n nodes. This is also evident from numerical solution using the spreadsheet, watching how the wave function behaves at large x as the energy is cranked up.The spreadsheet can also be used to plot the wave function for large n, say n = 200. It is instructive to compare the probability distribution with that for a *classical* pendulum, one oscillating with fixed amplitude and observed many times at random intervals. For the pendulum, the probability peaks at the end of the swing, where the pendulum is slowest and therefore spends most time. The $n = 200$ distribution amplitude follows this pattern, but of course oscillates. However, in the large n limit these oscillations take place over undetectably small intervals.The *classical* pendulum when not at rest clearly has a time-dependent probability distribution—it swings backwards and forwards. This means it *cannot* be in an eigenstate of the energy. In fact, the quantum state most like the classical is a *coherent state* built up of neighbouring energy eigenstates.

OPERATOR APPROACH TO THE SIMPLE HARMONIC OSCILLATOR

Having scaled the position coordinate x to the dimensionless ξ by $\xi = x/b = x\sqrt{m\omega/\hbar}$, let us also scale the momentum from p to $\pi = -id/d\xi$ (so $\pi = bp/\hbar = p/\sqrt{\hbar m\omega}$).

The Hamiltonian is

$$H = \frac{p^2 + m^2\omega^2x^2}{2m} = \frac{\hbar\omega}{2}\left(\pi^2 + \xi^2\right).$$

Dirac had the brilliant idea of factorizing this expression: the obvious thought $\left(\xi^2 + \pi^2\right) = (\xi + i\pi)(\xi - i\pi)$ isn't quite right, because it fails to take account of the noncommutativity of the operators, but the symmetrical version

$$H = \frac{\hbar\omega}{4}\left[(\xi + i\pi)(\xi - i\pi) + (\xi - i\pi)(\xi + i\pi)\right]$$

is fine, and we shall soon see that it leads to a very easy way of finding the eigenvalues and operator matrix elements for the oscillator, far simpler than using the wave functions we found above. Interestingly, Dirac's factorization here of a second-order differential operator into a product of first-order operators is close to the idea that led to his most famous achievement, the Dirac equation, the basis of the relativistic theory of electrons, protons, etc.

To continue, we define new operators $a, a^\dagger$ by

$$a = \frac{\xi + i\pi}{\sqrt{2}} = \frac{1}{\sqrt{2\hbar m\omega}}(m\omega x + ip), \quad a^\dagger = \frac{\xi - i\pi}{\sqrt{2}} = \frac{1}{\sqrt{2\hbar m\omega}}(m\omega x - ip).$$

From the commutation relation $\left[i\pi, \xi\right] = 1$ it follows that $[a, a^\dagger] = 1$. Therefore the Hamiltonian can be written:

$$H = \hbar\omega\left(a^\dagger a + \frac{1}{2}\right) = \hbar\omega\left(N + \frac{1}{2}\right), \text{ where } N = a^\dagger a.$$

Note that the operator N can only have *non-negative* eigenvalues, since

$$\langle\psi|N|\psi\rangle = \langle\psi|a^\dagger a|\psi\rangle = \langle\psi_a|\psi_a\rangle \geq 0.$$

Now

$$\left[N,a^\dagger\right]=a^\dagger aa^\dagger-a^\dagger a^\dagger a=a^\dagger\left[a,a^\dagger\right]=a^\dagger$$

Suppose N has an eigenfunction $|\nu\rangle$ with eigenvalue ν,

$N|\nu\rangle=\nu|\nu\rangle.$

From the two equations above

$Na^\dagger|\nu\rangle=a^\dagger N|\nu\rangle+a^\dagger|\nu\rangle=(\nu+1)a^\dagger|\nu\rangle$

so $a^\dagger|\nu\rangle$ is an eigenfunction of N with eigenvalue $\nu+1$. Operating with $a^\dagger$ again and again, we climb an infinite ladder of eigenstates equally spaced in energy. is often termed a creation operator, since the quantum of energy added each time it operates is equivalent to an added photon in black body radiation (electromagnetic oscillations in a cavity). It is easy to check that the state $a|\nu\rangle$ is an eigenstate with eigenvalue $\nu-1$, provided it is nonzero, so the operator a takes us *down* the ladder. However, this cannot go on indefinitely—we have established that N cannot have negative eigenvalues. We must eventually reach a state $|\nu\rangle$ for which $a|\nu\rangle=0$, the operator a *annihilates* the state. (At each step down, a annihilates one quantum of energy—so a is often called an *annihilation* or *destruction* operator.)

Since the norm squared of $a|\nu\rangle$, $\left|a|\nu\rangle\right|^2=\langle\nu|a^\dagger a|\nu\rangle=\langle\nu|N|\nu\rangle=\nu\langle\nu|\nu\rangle$, and since $\langle\nu|\nu\rangle>0$ for any nonvanishing state, it must be that *the lowest eigenstate* (the $|\nu\rangle$ for which $a|\nu\rangle=0$) *has* $\nu=0$. It follows that the ν's on the ladder are *the positive integers*, so from this point on we relabel the eigenstates with n in place of ν.

That is to say, we have proved that the only possible eigenvalues of N are zero and the positive integers: 0, 1, 2, 3... . N is called the *number operator*: it measures the number of quanta of energy in the oscillator above the irreducible ground state energy (that is, above the "zero-point energy" arising from the wave-like nature of the particle).

Since from above the Hamiltonian

$$H = \hbar\omega\left(a^\dagger a + \frac{1}{2}\right) = \hbar\omega\left(N + \frac{1}{2}\right)$$

the energy eigenvalues are

$$H|n\rangle = \left(n + \tfrac{1}{2}\right)\hbar\omega|n\rangle.$$

It is important to appreciate that Dirac's factorization trick and very little effort has given us *all* the eigenvalues of the Hamiltonian

$$H = \frac{\hbar\omega}{2}\left(\pi^2 + \xi^2\right).$$

Contrast the work needed in this section with that in the standard Schrodinger approach. We have also established that the lowest energy state $|0\rangle$, having energy $\frac{1}{2}\hbar\omega$, must satisfy the first-order differential equation $a|0\rangle = 0$, that is,

$$(\xi + i\pi)\,|\,0> = \left(\xi + \frac{d}{d\xi}\right)\psi_0(\xi) = 0.$$

The solution, unnormalized, is

$$\psi_0(\xi) = Ce^{-\xi^2/2}.$$

(In fact, we've seen this equation and its solution before: this was the condition for the "least uncertain" wave function in the discussion of the Generalized Uncertainty Principle.)

We denote the *normalized* set of eigenstates $|0\rangle, |1\rangle, |2\rangle, \ldots |n\rangle \ldots$ with $\langle n|n\rangle = 1$. Now $a^\dagger|n\rangle = C_n|n+1\rangle$, and C_n is easily found:

$$|C_n|^2 = |C_n|^2\langle n+1|n+1\rangle = \langle n|aa^\dagger|n\rangle = (n+1),$$

and

$$a^1|n\rangle = \sqrt{n+1}|n+1\rangle.$$

Therefore, if we take the set of orthonormal states $|0\rangle, |1\rangle, |2\rangle, \dots |n\rangle \dots$ as the basis in the Hilbert space, the *only* nonzero matrix elements of $a^\dagger$ are $\langle n+1|a^\dagger|n\rangle = \sqrt{n+1}$. That is to say,

$$a^\dagger = \begin{pmatrix} 0 & 0 & 0 & 0 & \dots \\ \sqrt{1} & 0 & 0 & 0 & \dots \\ 0 & \sqrt{2} & 0 & 0 & \dots \\ 0 & 0 & \sqrt{3} & 0 & \dots \\ \vdots & \vdots & \vdots & \vdots & \ddots \end{pmatrix}.$$

(The column vectors in the space this matrix operates on have an infinite number of elements: the lowest energy, the ground state component, is the entry at the *top* of the infinite vector—so up the energy ladder is down the vector!)

The adjoint

$$a = \begin{pmatrix} 0 & \sqrt{1} & 0 & 0 & \dots \\ 0 & 0 & \sqrt{2} & 0 & \dots \\ 0 & 0 & 0 & \sqrt{3} & \dots \\ 0 & 0 & 0 & 0 & \dots \\ \vdots & \vdots & \vdots & \vdots & \ddots \end{pmatrix}.$$

So

$$a|n\rangle = \sqrt{n}|n-1\rangle$$

For practical computations, we need to find the matrix elements of the position and momentum variables between the normalized eigenstates. Now

$$x = \sqrt{\hbar / 2m\omega}\left(a^\dagger + a\right), \quad p = i\sqrt{m\omega\hbar / 2}\left(a^\dagger - a\right)$$

so

$$x=\sqrt{\hbar/2m\omega}\begin{pmatrix}0 & \sqrt{1} & 0 & 0 & \cdots\\ \sqrt{1} & 0 & \sqrt{2} & 0 & \cdots\\ 0 & \sqrt{2} & 0 & \sqrt{3} & \cdots\\ 0 & 0 & \sqrt{3} & 0 & \cdots\\ \vdots & \vdots & \vdots & \vdots & \ddots\end{pmatrix},\quad p=i\sqrt{m\omega\hbar/2}\begin{pmatrix}0 & -\sqrt{1} & 0 & 0 & \cdots\\ \sqrt{1} & 0 & -\sqrt{2} & 0 & \cdots\\ 0 & \sqrt{2} & 0 & -\sqrt{3} & \cdots\\ 0 & 0 & \sqrt{3} & 0 & \cdots\\ \vdots & \vdots & \vdots & \vdots & \ddots\end{pmatrix}.$$

These matrices are, of course, Hermitian (not forgetting the i factor in p).

To find the matrix elements between eigenstates of any product of x's and p's, express all the x's and p's in terms of a's and $a^\dagger$'s, to give a sum of products of a's and $a^\dagger$'s. Each product in this sum can be evaluated sequentially from the right, because each a or $a^\dagger$ has only one nonzero matrix element when the product operates on one eigenstate.

NORMALIZING THE EIGENSTATES IN X-SPACE

The normalized ground state wave function is

$$\psi_0(\xi)=Ce^{-\xi^2/2}=\left(\frac{m\omega}{\pi\hbar}\right)^{1/4}e^{-m\omega x^2/2\hbar},$$

where we have gone back to the x variable, and normalized using $\int_{-\infty}^{\infty}e^{-ax^2}dx=\sqrt{\pi/a}$.

To find the normalized wave functions for the higher states, they are first constructed formally by applying the creation operator $a^\dagger$ repeatedly on the ground state $|0\rangle$. Next, the result is translated into x-space (actually $\xi=x/b$) by writing $a^\dagger$ as a differential operator, acting on $\psi_0(\xi)$.

Using $\langle n|a^\dagger|n-1\rangle=\sqrt{n}$,

$$|n\rangle=\frac{a^\dagger}{\sqrt{n}}|n-1\rangle=\ldots=\frac{(a^\dagger)^n}{\sqrt{n!}}|0\rangle.$$

Now

$$a^\dagger = \left(1/\sqrt{2}\right)(\xi - i\pi) = \left(1/\sqrt{2}\right)(\xi - d/d\xi),$$

so

$$\psi_n(\xi) = \frac{\left(a^\dagger\right)^n}{\sqrt{n!}}|0\rangle = \frac{1}{\sqrt{n!}}\left(\frac{1}{\sqrt{2}}\left(\xi - \frac{d}{d\xi}\right)\right)^n \left(\frac{m\omega}{\pi\hbar}\right)^{1/4} e^{-\xi^2/2}.$$

We need to check that this expression is indeed the same as the Hermite polynomial wave function derived earlier, and to do that we need some further properties of the Hermite polynomials.

SOME PROPERTIES OF HERMITE POLYNOMIALS

The mathematicians *define* the Hermite polynomials by:

$$H_n(\xi) = (-)^n e^{\xi^2} \frac{d^n}{d\xi^n} e^{-\xi^2}$$

so

$$H_0(\xi) = 1,\ H_1(\xi) = 2\xi,\ H_2(\xi) = 4\xi^2 - 2,\ H_3(\xi) = 8\xi^3 - 12\xi, \text{ etc.}$$

It follows immediately from the definition that the coefficient of the leading power is 2^n.

It is a straightforward exercise to check that H_n is a solution of the differential equation

$$\left(\frac{d^2}{d\xi^2} - 2\xi\frac{d}{d\xi} + 2n\right)H_n(\xi) = 0,$$

so these are indeed the same polynomials we found by the series solution of Schrodinger's equation earlier (recall the equation for the polynomial component of the wave function was

$$\frac{d^2h}{d\xi^2} - 2\xi\frac{dh}{d\xi} + (2\varepsilon - 1)h = 0,$$

with $2\varepsilon = 2n+1$.)

We have found $\psi_n(\xi)$ in the form

$$\psi_n(\xi) = \frac{1}{\sqrt{n!}}\left(\frac{1}{\sqrt{2}}\left(\xi - \frac{d}{d\xi}\right)\right)^n \left(\frac{m\omega}{\pi\hbar}\right)^{1/4} e^{-\xi^2/2}.$$

We shall now prove that the polynomial component is exactly equivalent to the Hermite polynomial as defined at the beginning of this section.

We begin with the operator identity:

$$\left(\xi - \frac{d}{d\xi}\right) = -e^{\xi^2/2}\frac{d}{d\xi}e^{-\xi^2/2}$$

Both sides of this expression are to be regarded as *operators*, that is, it is assumed that both are operating on some function $f(\xi)$.

Now take the nth power of both sides: on the right, we find, for example,

$$\left(-e^{\xi^2/2}\frac{d}{d\xi}e^{-\xi^2/2}\right)^3 = (-)^3 e^{\xi^2/2}\frac{d}{d\xi}e^{-\xi^2/2}e^{\xi^2/2}\frac{d}{d\xi}e^{-\xi^2/2}e^{\xi^2/2}\frac{d}{d\xi}e^{-\xi^2/2}$$

$$= (-)^3 e^{\xi^2/2}\frac{d^3}{d\xi^3}e^{-\xi^2/2}$$

since the intermediate exponential terms cancel against each other.

So:

$$\left(\xi - \frac{d}{d\xi}\right)^n = (-)^n e^{\xi^2/2}\frac{d^n}{d\xi^n}e^{-\xi^2/2}$$

and substituting this into the expression for $\psi_n(\xi)$ above,

$$\psi_n(\xi) = \frac{1}{\sqrt{2^n n!}}(-)^n\left(e^{\xi^2/2}\frac{d^n}{d\xi^n}e^{-\xi^2/2}\right)\left(\frac{m\omega}{\pi\hbar}\right)^{1/4} e^{-\xi^2/2}$$

$$= \frac{1}{\sqrt{2^n n!}}(-)^n \left(\frac{m\omega}{\pi\hbar}\right)^{1/4} e^{-\xi^2/2}\left(e^{\xi^2}\frac{d^n}{d\xi^n}e^{-\xi^2}\right)$$

$$= \frac{1}{\sqrt{2^n n!}}\left(\frac{m\omega}{\pi\hbar}\right)^{1/4} H_n(\xi)e^{-\xi^2/2}, \text{ with } \xi = \sqrt{\frac{m\omega}{\hbar}}x.$$

This established the equivalence of the two approaches to Schrodinger's equation for the simple harmonic oscillator, and provides us with the overall normalization constants without doing integrals. (The expression for $\psi_n(\xi)$ above satisfies $\int |\psi_n|^2\, dx = 1$.)

Use $H_n(\xi) = (-)^n e^{\xi^2}\frac{d^n}{d\xi^n}e^{-\xi^2}$ to prove:

(a) the coefficient of ξ^n is 2^n.

(b) $H_n'(\xi) = 2nH_{n-1}(\xi)$

(c) $H_{n+1}(\xi) = 2\xi H_n(\xi) - 2nH_{n-1}(\xi)$

(d) $\int_{-\infty}^{\infty} e^{-\xi^2}H_n^2(\xi)d\xi = 2^n n!\sqrt{\pi}$

(Hint: rewrite as $\int_{-\infty}^{\infty} H_n(\xi)(-)^n\frac{d^n}{d\xi^n}e^{-\xi^2}d\xi$, then integrate by parts n times, and use (a).)

(e) $\int_{-\infty}^{\infty} e^{-\xi^2}H_n(\xi)H_m(\xi)d\xi = 0$, for $m \neq n$.

It's worth doing these exercises to become more familiar with the Hermite polynomials, but in evaluating matrix elements (and indeed in establishing some of these results) it is almost always far simpler to work with the creation and annihilation operators.

Use the creation and annihilation operators to find $\langle n|x^4|n\rangle$. This matrix element is useful in estimating the energy change arising on adding a small nonharmonic potential energy term to a harmonic oscillator.

Chapter 13

Time-Dependent Wave Functions

The set of normalized eigenstates $|0\rangle, |1\rangle, |2\rangle,$ $\ldots|n\rangle\ldots$ discussed above are of course solutions to the time-*independent* Schrodinger equation, or in ket notation eigenstates of the Hamiltonian $H|n\rangle = (n+\frac{1}{2})\hbar\omega|n\rangle$. Putting in the time-dependence explicitly, $|n,t\rangle = e^{-iHt/\hbar}|n,t=0\rangle = e^{-i(n+\frac{1}{2})\omega t}|n\rangle$. It is necessary to include the time dependence when dealing with a state which is a superposition of states of different energies, such as $(1/\sqrt{2})(|0\rangle+|1\rangle)$, which then becomes $(1/\sqrt{2})(e^{-i\omega t/2}|0\rangle+e^{-3i\omega t/2}|1\rangle)$. Expectation values of combinations of position and/or momentum operators in such states are best evaluated by expressing everything in terms of annihilation and creation operators.

SOLVING SCHRODINGER'S EQUATION IN MOMENTUM SPACE

In the lecture on Function Spaces, we established that the basis of $|x\rangle$ states (eigenstates of the position operator) and that of $|k\rangle$ states (eigenstates of the momentum operator) were both complete bases in Hilbert space (physicist's definition) so we could work equally well with either from a formal point of view. Why then do we almost always work in x-space? Well, probably because we live in x-space, but there's another reason. The momentum operator in the x-space representation is $p = -i\hbar d/dx$, so Schrodinger's equation, written $\left(p^2/2m + V(x)\right)\psi(x) = E\psi(x)$, with p in operator form, is a second-order differential equation. Now consider what happens to Schrodinger's equation if we work in p-space. Since the operator identity $[x,p] = i\hbar$ is true regardless of representation, we must have $x = i\hbar d/dp$. So for a particle in a potential $V(x)$, writing Schrodinger's equation in p-space we are confronted with the nasty looking operator $V(i\hbar d/dp)$! This will produce a differential equation in general a lot harder to solve than the standard x-space equation—so we stay in x-space.

But there are two potentials that can be handled in momentum space: first, for a *linear* potential $V(x) = -Fx$, the momentum space analysis is actually easier—it's just a first-order equation. Second, for a particle in a *quadratic* potential—a simple harmonic oscillator—the two approaches yield the *same* differential equation. That means that the eigenfunctions in momentum space (scaled appropriately) must be *identical* to those in position space—the simple harmonic eigenfunctions are their own Fourier transforms!

Chapter 14

Time-Dependent Solutions: Propagators and Representations

We've spent most of the course so far concentrating on the eigenstates of the Hamiltonian, states whose time-dependence is merely a changing phase. We did mention much earlier a superposition of two different energy states in an infinite well, resulting in a wave function sloshing backwards and forwards. It's now time to cast the analysis of time dependent states into the language of bras, kets and operators. We'll take a time-independent Hamiltonian *H*, with a complete set of orthonormalized eigenstates, and as usual

$$i\hbar\frac{\partial\psi(x,t)}{\partial t}=-\frac{\hbar^2}{2m}\frac{\partial^2\psi(x,t)}{\partial x^2}+V(x)\psi(x,t),$$

Or, as we would now write it

$$i\hbar\frac{\partial}{\partial t}\left|\psi(x,t)\right\rangle=H\left|\psi(x,t)\right\rangle.$$

Since *H* is itself time independent, this is very easy to integrate!

$$\left|\psi(x,t)\right\rangle=e^{-iH(t-t_0)/\hbar}\left|\psi(x,t_0)\right\rangle.$$

The exponential operator that generates the time-dependence is called the *propagator*, because it describes how the wave propagates from its initial configuration, and is usually denoted by *U*:

$$|\psi(x,t)\rangle = U(t-t_0)|\psi(x,t_0)\rangle.$$

It's appropriate to call the propagator U, because it's a unitary operator:

$$U(t-t_0) = e^{-iH(t-t_0)/\hbar}, \text{ so } U^\dagger(t-t_0) = e^{iH^\dagger(t-t_0)/\hbar} = e^{iH(t-t_0)/\hbar} = U^{-1}(t-t_0).$$

Since H is hermitian, U is unitary. It immediately follows that

$$\langle\psi(x,t)|\psi(x,t)\rangle = \langle\psi(x,t_0)|U^\dagger U(t-t_0)|\psi(x,t_0)\rangle = \langle\psi(x,t_0)|\psi(x,t_0)\rangle$$

the norm of the ket vector is conserved, or, translating to wave function language, a wave function correctly normalized to give a total probability of one stays that way. (This can also be proved from the Schrodinger equation, of course, but this is quicker.)

This is all very succinct, but unfortunately the exponential of a second-order differential operator doesn't sound too easy to work with. Recall, though, that any function of a Hermitian operator has the same set of eigenstates as the original operator. This means that the eigenstates of $e^{-iH(t-t_0)/\hbar}$ are the *same* as the eigenstates of H, and if $H|\psi_n\rangle = E_n|\psi_n\rangle$, then

$$e^{-iH(t-t_0)/\hbar}|\psi_n\rangle = e^{-iE_n(t-t_0)/\hbar}|\psi_n\rangle.$$

This is of course nothing but the time dependent phase factor for the eigenstates we found before—and, as before, to find the time dependence of any general state we must express it as a superposition of these eigenkets, each having its own time dependence. But how do we do that in the operator language? Easy: we simply insert an identity operator, the one constructed from the complete set of eigenkets, thus:

$$|\psi(t)\rangle = e^{-iH(t-t_0)/\hbar}\sum_{n=1}^{\infty}|\psi_n\rangle\langle\psi_n|\psi(t_0)\rangle = \sum_{n=1}^{\infty} e^{-iE_n(t-t_0)/\hbar}|\psi_n\rangle\langle\psi_n|\psi(t_0)\rangle.$$

Staring at this, we see that it's just what we had before: at the initial time $t = t_0$ the wave function can be written as a sum over the eigenkets:

$$|\psi(t_0)\rangle = \sum |\psi_n(t_0)\rangle\langle\psi_n(t_0)|\psi(t_0)\rangle = \sum c_n|\psi_n(t_0)\rangle$$

with $c_n = \langle\psi_n|\psi\rangle$, $\sum|c_n|^2 = 1$, and the usual generalization for continuum eigenvalues, and the time development is just given by inserting the phases:

$$|\psi(t)\rangle = \sum c_n e^{-iE_n(t-t_0)/\hbar}|\psi_n(t_0)\rangle$$

The expectation value of the energy E in $|\psi\rangle$,

$$\langle E\rangle = \langle\psi|H|\psi\rangle = \sum|c_n|^2 E_n$$

and is (of course) time independent.

The expectation value of the particle position x is

$$\langle\psi(t)|x|\psi(t)\rangle = \sum_{n,m} c_n^* c_m e^{i(E_n-E_m)(t-t_0)/\hbar}\langle\psi_n(t_0)|x|\psi_m(t_0)\rangle$$

and is *not* in general time-independent. (It *is* real, of course, on adding the (n,m) term to the (m,n) term.) This analysis is only valid for a *time-independent* Hamiltonian. The important extension to a system in a time-dependent external field, such as an atom in a light beam.

THE FREE PARTICLE PROPAGATOR

To gain some insight into what the propagator U looks like, we'll first analyse the case of a particle in one dimension with no potential at all.

We'll also take $t_0=0$ to make the equations less cumbersome.

For a free particle in one dimension $E = p^2/2m = \hbar^2k^2/2m$ the energy eigenstates are also momentum eigenstates, we label them $|k\rangle$, so

$$U(t) = e^{-iHt/\hbar} = \int_{-\infty}^{\infty} e^{-iHt/\hbar}\frac{dk}{2\pi}|k\rangle\langle k| = \int_{-\infty}^{\infty} e^{-i\hbar k^2 t/2m}\frac{dk}{2\pi}|k\rangle\langle k|$$

Let's consider (following Shankar and others) what seems the simplest example.

Suppose that at $t = t_0 = 0$, a particle is at x_0: $\psi(x,t=0)=\delta(x-x_0)=|x_0\rangle$: what is the probability amplitude for finding it at x at a later time t?

$$\langle x|U(t,0)|x_0\rangle = \int_{-\infty}^{\infty} e^{-i\hbar k^2 t/2m}\frac{dk}{2\pi}\langle x|k\rangle\langle k|x_0\rangle$$

$$= \int_{-\infty}^{\infty} e^{-i\hbar k^2 t/2m}\frac{dk}{2\pi}e^{-ik(x_0-x)}$$

$$= \sqrt{\frac{m}{2\pi\hbar it}}e^{im(x_0-x)^2/2\hbar t},$$

using the standard identity for Gaussian integrals,

$$\int_{-\infty}^{\infty} dk e^{-ak^2+bk} = \sqrt{\frac{\pi}{a}}e^{b^2/4a}.$$

On examining the above expression, though, it turns out to be nonsense! Noting that the term in the exponent is pure imaginary, $|\psi(x,t)|^2 = m/2\pi\hbar t$ independent of x! This particle apparently instantaneously fills all of space, but then its probability dies away as $1/t$

Notice first that $|\psi(x,t)|^2$ is constant throughout space. This means that the normalization, $\int|\psi(x,t)|^2 dx = \infty$! And, as we've seen above, the normalization stays constant in time—the propagator is unitary. Therefore, our initial wave function must have had infinite norm. That's exactly right—we took the initial wave function $\psi(x,t=0)=\delta(x-x_0)=|x_0\rangle$.

Think of the δ-function as a limit of a function equal to $1/\Delta$ over an interval of length Δ, with Δ going to zero, and it's clear the normalization goes to infinity as $1/\Delta$. This is not a meaningful wave function for a particle. Recall that continuum kets like $|x_0\rangle$ are normalized by

$\langle x|x'\rangle = \delta(x-x')$, they do *not* represent wave functions individually normalizable in the usual sense. The only meaningful wave functions are *integrals over a range* of such kets, such as $\int dx\psi(x)|x\rangle$. In an integral like this, notice that states $|x\rangle$ within some tiny x-interval of length δx, say, have total weight $\psi(x)\,\delta x$, which goes to zero as δx is made smaller, but by writing $\psi(x,t=0)=\delta(x-x_0)=|x_0\rangle$ we took a single such state and gave it a finite weight. This we can't do.

Of course, we do want to know how a wave function initially localized *near* a point develops. To find out, we must apply the propagator to a legitimate wave function—one that is normalizable to begin with. The simplest "localized particle" wave function from a practical point of view is a Gaussian wave packet,

$$\psi(x',0) = e^{ip_0x'/\hbar}\frac{e^{-x'^2/2d^2}}{(\pi d^2)^{1/4}}.$$

We used d in place of Δ here

The wave function at a later time is then given by the operation of the propagator on this initial wave function:

$$\psi(x,t) = \int U(x,t;x',0)e^{ip_0x'/\hbar}\frac{e^{-x'^2/2d^2}}{(\pi d^2)^{1/4}}dx' =$$

$$\sqrt{\frac{m}{2\pi\hbar i t}}\int e^{im(x-x')^2/2\hbar t}e^{ip_0x'/\hbar}\frac{e^{-x'^2/2d^2}}{(\pi d^2)^{1/4}}dx'.$$

Note first that since this is just $|\psi(x,t)\rangle = U(t)|\psi(x,t=0)\rangle$ written explicitly in terms of Schrodinger wave functions,

$$\psi(x,t) = \int U(x,t;x',0)\psi(x',0)\,dx'$$

It is evident that $U(x,t;x',0)\to\delta(x-x')$ as $t\to 0$. This is just equivalent to the operator statement that $e^{-iHt/\hbar}\to I$, the unit operator, as $t\to 0$.

The integral over $x2$ is just another Gaussian integral, so we use the same result,

$$\int_{-\infty}^{\infty} dx' e^{-ax'^2+bx'} = \sqrt{\frac{\pi}{a}}e^{b^2/4a}.$$

Looking at the expression above, we can see that

$$b=-\frac{im}{\hbar t}\left(x-\frac{p_0 t}{m}\right),\quad a=\frac{1}{2d^2}-\frac{im}{2\hbar t}.$$

This gives

$$\psi(x,t)=\frac{\pi^{-1/4}}{\sqrt{d\left(1+\frac{i\hbar t}{md^2}\right)}}\exp\left(\frac{imx^2}{2\hbar t}\right)\exp\left(-\frac{\left(\frac{im}{\hbar t}\left(x-\frac{p_0 t}{m}\right)\right)^2}{2\left(1+\frac{i\hbar t}{md^2}\right)}\right)$$

where the second exponential is the term $e^{b^2/4a}$. As written, the small t limit is not very apparent, but some algebraic rearrangement yields:

$$\psi(x,t)=\frac{\pi^{-1/4}}{\sqrt{d\left(1+i\hbar t/md^2\right)}}\exp\left(-\frac{\left(x-p_0 t/m\right)^2}{2d^2\left(1+i\hbar t/md^2\right)}\right)\exp\left(\frac{ip_0}{\hbar}\left(x-p_0 t/2m\right)\right).$$

Written this way, it is evident that the expression *goes to the initial wave packet* for t going to zero, as of course it must.

Although the phase in the above expression for $\psi(x,t)$ has contributions from all three terms, the main phase oscillation is in the third term, and one can see the phase velocity is one-half the group velocity, as discussed earlier.

The resulting *probability density*:

$$|\psi(x,t)|^2=\frac{1}{\sqrt{\pi\left(d^2+\hbar^2t^2/m^2d^2\right)}}\cdot\exp^{-\frac{(x-p_0t/m)^2}{(d^2+\hbar^2t^2/m^2d^2)}}.$$

This is a Gaussian wave packet, having a width which goes as $\hbar t/md$ for large times, where d is the width of the initial packet in x-space—so $\hbar/md$ is the spread in *velocities* $\langle\Delta v\rangle$ within the packet, hence the gradual spreading $\langle\Delta v\rangle t$ in x-space.

It's amusing to look at the limit of this as the width d of the initial Gaussian packet goes to zero, and see how that relates to our δ-function result. Suppose we are at distance x from the origin, and there is initially a Gaussian wave packet centered at the origin, width $d \ll x$. At time $t \sim mxd/\hbar$, the wave packet has spread to x and has $|\psi(x,t)|^2$ of order $1/x$ at x. Thereafter, it continues to spread at a linear rate in time, so locally $|\psi(x,t)|^2$ must decrease as $1/t$ to conserve probability. In the δ-function limit $d \to 0$, the wave function instantly spreads through a huge volume, but then goes as $1/t$ as it spreads into an even huger volume. Or something.

SCHRODINGER AND HEISENBERG REPRESENTATIONS

Assuming a Hamiltonian with no explicit time dependence, the time-dependent Schrodinger equation has the form

$$i\hbar\frac{\partial}{\partial t}|\psi(x,t)\rangle = H|\psi(x,t)\rangle$$

and as discussed above, the formal solution can be expressed as:

$$|\psi(x,t)\rangle = e^{-iHt/\hbar}|\psi(x,t=0)\rangle.$$

Now, any measurement on a system amounts to measuring a matrix element of an operator between two states (or, more generally, a function of such matrix elements).

In other words, the *physically significant time dependent quantities* are of the form

$$\langle\varphi(t)|A|\psi(t)\rangle = \langle\varphi(0)|e^{iHt/\hbar}Ae^{-iHt/\hbar}|\psi(0)\rangle$$

where A is an operator, which we are assuming has no explicit time dependence.

So in this Schrodinger picture, the time dependence of the measured value of an operator like x or p comes about because we measure the matrix element of an unchanging operator between bras and kets that are changing in time.

Heisenberg took a different approach: he assumed that the ket describing a quantum system did *not* change in time, it remained at $|\psi(0)\rangle$, but the *operators* evolved according to:

$$A_H(t) = e^{iHt/\hbar}A_H(0)e^{-iHt/\hbar}.$$

Clearly, this leads to the same physics as before. The equation of motion of the operator is:

$$i\hbar\frac{dA_H(t)}{dt} = [A_H(t), H].$$

The Hamiltonian itself does not change in time—energy is conserved, or, to put it another way, H commutes with $e^{-iHt/\hbar}$. But for a nontrivial Hamiltonian, say for a particle in one dimension in a potential,

$$H = p^2/2m + V(x)$$

the separate components *will* have time-dependence, parallel to the classical case: the *kinetic* energy of a swinging pendulum varies with time. (For a particle in a potential in an energy eigenstate the expectation value of the kinetic energy is constant, but this is not the case for any other state, that is, for a superposition of different eigenstates.) Nevertheless, the *commutator* of x, p will be time-independent:

$$[x_H(t), p_H(t)] = e^{iHt/\hbar}[x_H(0), p_H(0)]e^{-iHt/\hbar} = e^{iHt/\hbar}i\hbar e^{-iHt/\hbar} = i\hbar$$

(The Heisenberg operators are identical to the Schrodinger operators at $t = 0$.)

Applying the general commutator result $[A,BC]=[A,B]C+B[A,C]$,

$$\left[x_H(t),\frac{p^2{}_H(t)}{2m}\right]=\frac{i\hbar p_H(t)}{m}$$

so

$$\frac{dx_H(t)}{dt}=\frac{p_H(t)}{m}$$

and since $[x_H(t),p_H(t)]=i\hbar,\ p_H(t)=-i\hbar d/dx_H(t)$,

$$\frac{dp_H(t)}{dt}=\frac{1}{i\hbar}\left[p_H(t),V(x_H(t))\right]=-\nabla V(x_H(t)).$$

This result could also be derived by writing $V(x)$ as an expansion in powers of x, then taking the commutator with p.

Notice from the above equations that the operators in the Heisenberg Representation obey the *classical* laws of motion! *Ehrenfest's Theorem*, that the *expectation values* of operators in a quantum state follow the classical laws of motion, follows immediately, by taking the expectation value of both sides of the operator equation of motion in a quantum state.

SIMPLE HARMONIC OSCILLATOR IN THE HEISENBERG REPRESENTATION

For the simple harmonic oscillator, the equations are easily integrated to give:

$$x_H(t)=x_H(0)\cos\omega t+(p_H(0)/m\omega)\sin\omega t$$

$$p_H(t)=p_H(0)\cos\omega t-m\omega x_H(0)\sin\omega t.$$

We have put in the $_H$ subscript to emphasize that these are operators. It is usually clear from the context that the Heisenberg representation is being used, and the subscript $_H$ may be safely omitted.

The time-dependence of the *annihilation operator* a is:

$$a(t) = e^{iHt/\hbar} a(0) e^{-iHt/\hbar}$$

with

$$H = \hbar\omega\left(a^\dagger(t)a(t) + \tfrac{1}{2}\right).$$

Note again that although H is itself time-independent, it is necessary to include the time-dependence of individual operators within H.

$$\begin{aligned} i\hbar\frac{d}{dt}a(t) &= \left[a(t), H\right] = \hbar\omega\left[a(t), a^\dagger(t)a(t)\right] \\ &= \hbar\omega\left[a(t), a^\dagger(t)\right]a(t) = \hbar\omega a(t) \end{aligned}$$

so

$$a(t) = a(0)e^{-i\omega t}.$$

Actually, we could have seen this as follows: if $|n\rangle$ are the energy eigenstates of the simple harmonic oscillator,

$$e^{-iHt/\hbar}|n\rangle = e^{-in\hbar\omega t/\hbar}|n\rangle = e^{-in\omega t}|n\rangle.$$

Now the *only* nonzero matrix elements of the annihilation operator $\hat{a}$ between energy eigenstates are of the form

$$\begin{aligned} \langle n-1|a(t)|n\rangle = \langle n-1|e^{iHt/\hbar}a(0)e^{-iHt/\hbar}|n\rangle &= e^{i\omega(n-1)t}\langle n-1|a(0)|n\rangle e^{-i\omega nt} \\ &= \langle n-1|a(0)|n\rangle e^{-i\omega t} \end{aligned}$$

Since this time-dependence is true of *all* energy matrix elements (trivially so for most of them, since they're identically zero), and the eigenstates of the Hamiltonian span the space, it is true as an *operator* equation.

Evidently, the expectation value of the operator $a(t)$ in any state goes clockwise in a circle centered at the origin in the complex plane. That this is indeed the classical motion of the simple harmonic oscillator is confirmed by recalling the definition $a = \frac{\xi + i\pi}{\sqrt{2}} = \frac{1}{\sqrt{2\hbar m\omega}}(m\omega x + ip)$, so the complex plane corresponds to the $(m\omega x, p)$ phase space discussed near the beginning of the lecture on the Simple Harmonic Oscillator. We'll discuss this in much more detail in the next lecture, on Coherent States.

The time-dependence of the creation operator is just the adjoint equation: $a^\dagger(t) = a^\dagger(0)e^{i\omega t}$.

Chapter 15

Coherent States of the Simple Harmonic Oscillator

Consider a macroscopic simple harmonic oscillator, and to keep things simple assume there are no interactions with the rest of the universe. We know how to describe the motion using classical mechanics: for a given initial position and momentum, classical mechanics correctly predicts the future path, as confirmed by experiments with real (admittedly not perfect) systems. But from the Hamiltonian we could also write down Schrodinger's equation, and from that predict the future behaviour of the system. Since we already know the answer from classical mechanics and experiment, quantum mechanics must give us the same result in the limiting case of a large system.

It is a worthwhile exercise to see just how this happens. Evidently, we cannot simply follow the classical method of specifying the initial position and momentum—the uncertainty principle won't allow it. What we can do, though, is to take an initial state in which the position and momentum are specified *as precisely as possible*. Such a state is called a *minimum uncertainty* state (the details can be found in my earlier lecture on the Generalized Uncertainty Principle).

In fact, the *ground state* of a simple harmonic oscillator *is* a minimum uncertainty state. This is not too surprising—it's just a localized wave packet centered at the origin. The system is as close to rest as possible, having only zero-point motion.

What *is* surprising is that there are excited states of the pendulum in which this ground state wave packet swings backwards and forwards indefinitely, a quantum realization of the classical system, and the wave packet is always one of minimum uncertainty. Recall that this *doesn't* happen for a *free* particle on a line—in that case, an initial minimal uncertainty wave packet spreads out because the different momentum components move at different speeds. But for the oscillator, the potential somehow keeps the wave packet together, a minimum uncertainty wave packet at all times. These remarkable quasi-classical states are called *coherent states,* and were discovered by Schrodinger himself. They are important in many quasi-classical contexts, including laser radiation.

Our task here is to construct and analyse these coherent states and to find how they relate to the usual energy eigenstates of the oscillator.

CLASSICAL MECHANICS OF THE SIMPLE HARMONIC OSCILLATOR

To define the notation, let us briefly recap the dynamics of the *classical* oscillator: the constant energy is

$$E = \frac{p^2}{2m} + \frac{1}{2}kx^2$$

or

$$p^2 + (m\omega x)^2 = 2mE, \quad \omega = \sqrt{k/m}.$$

The classical motion is most simply described in *phase space,* a two-dimensional plot in the variables $(m\omega x, p)$. In this space, the point $(m\omega x, p)$ corresponding to the position and momentum of the oscillator at an instant of time moves as time progresses at constant angular speed ω in a clockwise direction around the circle of radius $\sqrt{2mE}$ centered at the origin.

(*Note*: phase space is usually defined in terms of the variables (x, p), but in describing the simple harmonic oscillator, the pair $(m\omega x, p)$ are more convenient.)

This motion is elegantly described by regarding the two-dimensional phase space as a complex plane, and defining the dimensionless complex variable

$$z = \frac{m\omega x + ip}{\sqrt{2\hbar m\omega}}.$$

The time evolution in phase space is simply $z(t) = z_0 e^{-i\omega t}$.

The particular choice of (quantum!) scaling factor in defining z amounts to defining the unit of energy as $\hbar\omega$, the natural unit for the oscillator: it is easy to check that if the classical energy $E = \left(n + \frac{1}{2}\right)\hbar\omega$ then the dimensionless $|z|^2$ is simply the number $n + \frac{1}{2}$ (which is of course very large, so the ½ is insignificant).

MINIMUM UNCERTAINTY WAVE PACKETS

We established in the lecture on the Generalized Uncertainty Principle that any minimum uncertainty one-dimensional wave function (so $\Delta p \cdot \Delta x = \hbar/2$) for a particle must satisfy the linear differential equation (here $\hat{p} = -i\hbar d/dx$)

$$\left(\hat{p} - \langle p \rangle\right)\psi(x) = \lambda\left(\hat{x} - \langle x \rangle\right)\psi(x)$$

where $\langle x \rangle, \langle p \rangle, \lambda$ are constants, and λ is *pure imaginary*. The equation is easy to solve: any minimum uncertainly one-dimensional wave function is a Gaussian wave packet, having expectation value of momentum $\langle p \rangle$, centered at $\langle x \rangle$ and having width $(\Delta x)^2 = -\hbar/2i\lambda$. ($\Delta x$ is defined for a state $|\psi\rangle$ by $(\Delta x)^2 = \langle \psi | \left(x - \langle x \rangle\right)^2 | \psi \rangle$.) That is to say, the minimum uncertai ıly solution is:

$$\psi(x) = Ce^{i\langle p\rangle x/\hbar}e^{i\lambda(x-\langle x\rangle)^2/2\hbar} = Ce^{i\langle p\rangle x/\hbar}e^{-(x-\langle x\rangle)^2/4(\Delta x)^2}$$

with C the normalization constant.

In fact, the simple harmonic oscillator *ground state* $\psi_0(x) = \left(\frac{m\omega}{\pi\hbar}\right)^{1/4} e^{-m\omega x^2/2\hbar}$ is just such a minimum uncertainty state, with

$$\lambda = im\omega,\ \langle x\rangle = \langle p\rangle = 0;\ (\Delta x)^2 = \frac{\hbar}{2m\omega},\ \ (\Delta p)^2 = \frac{\hbar m\omega}{2},\ \ \Delta p\cdot\Delta x = \frac{\hbar}{2}.$$

Furthermore, it is easy to see that the displaced ground state $\psi_0(x-\langle x_0\rangle) = Ce^{-m\omega(x-\langle x_0\rangle)^2/2\hbar}$ (writing the normalization constant $(m\omega/\pi\hbar)^{1/4} = C$) must also be a minimum uncertainty state, *with the same* $\lambda = im\omega$. Of course, in contrast to the ground state, this displaced state is no longer an eigenstate of the Hamiltonian, and any such initial state will change with time.

(Both these states have the same spread in x-space $(\Delta x)^2 = \hbar/2m\omega$, and the same spread in p-space, the only difference in the p direction being a phase factor $e^{ip\langle x_0\rangle/\hbar}$ for the displaced state.)

What about the higher eigenstates of the oscillator Hamiltonian? They are *not* minimally uncertain states—for the nth state, $\Delta p\cdot\Delta x = n\hbar/2$, as is easily checked using $\frac{1}{2}(\Delta p)^2/2m = \frac{1}{2}k(\Delta x)^2 \sim \frac{1}{2}n\hbar\omega$. So, if we construct a minimally uncertain higher energy state, it will *not* be an eigenstate of the Hamiltonian.

TIME DEVELOPMENT OF A COHERENT STATE: THE ROLE OF THE ANNIHILATION OPERATOR

In this section, we shall establish a remarkable connection between minimally uncertain oscillator states and the annihilation operator, then use properties of that operator to find the time-development of the minimally uncertain states.

Suppose that at $t = 0$ the oscillator wave function is the minimum uncertainty state

$$\psi(x,t=0) = Ce^{i\langle p_0\rangle x/\hbar}e^{i\lambda(x-\langle x_0\rangle)^2/2\hbar} = Ce^{i\langle p_0\rangle x/\hbar}e^{-m\omega(x-\langle x_0\rangle)^2/2\hbar}$$

centered at $\langle p_0\rangle, m\omega\langle x_0\rangle$ in phase space (as defined above for the classical oscillator), and with $\lambda = im\omega$ to give it the same spatial extent as the ground state.

From the preceding section, this $\psi(x,0)$ satisfies the minimum uncertainty equation

$$(\hat{p}-\langle p_0\rangle)\psi(x,0) = im\omega(\hat{x}-\langle x_0\rangle)\psi(x,0).$$

Rearranging this equation (and multiplying by $-i$) shows it in a different light:

$$(m\omega\hat{x}+i\hat{p})\psi(x,0) = (m\omega\langle x_0\rangle+i\langle p_0\rangle)\psi(x,0).$$

This is an eigenvalue equation! The wave packet $\psi(x,0)$ is an eigenstate of the operator $(m\omega\hat{x}+i\hat{p})$ with eigenvalue $(m\omega\langle x_0\rangle+i\langle p_0\rangle)$. It is *not*, of course, an eigenstates of either $\hat{p}$ or $\hat{x}$ taken individually.

Furthermore, the operator $(m\omega\hat{x}+i\hat{p})$ is just a constant times the annihilation operator $\hat{a}$ —recall

$$\hat{a} = \frac{1}{\sqrt{2\hbar m\omega}}(m\omega\hat{x}+i\hat{p}).$$

Therefore, this minimally uncertain initial wave packet $\psi(x,0)$ is an eigenstate of the annihilation operator $\hat{a}$, with eigenvalue $(m\omega\langle x_0\rangle+i\langle p_0\rangle)/\sqrt{2\hbar m\omega}$. By the way, it's ok for $\hat{a}$ to have complex eigenvalues, because $\hat{a}$ is not a hermitian operator.

We can now make the connection with the complex plane representation of the classical operator: the eigenvalue

$\left(m\omega\langle x_0\rangle+i\langle p_0\rangle\right)/\sqrt{2\hbar m\omega}$ is precisely the dimensionless complex parameter z_0!

This means that if we have a minimal uncertainty oscillator wave packet

$$\psi(x,t=0)=Ce^{i\langle p_0\rangle x/\hbar}e^{-m\omega\left(x-\langle x_0\rangle\right)^2/2\hbar}$$

having the same spatial extent as the ground state, centered at $m\omega\langle x_0\rangle,\langle p_0\rangle$ in phase space, and we write

$$z_0=\frac{m\omega\langle x_0\rangle+i\langle p_0\rangle}{\sqrt{2\hbar m\omega}}$$

then

$$\hat{a}\psi(x,t=0)=z_0\psi(x,t=0).$$

That is to say, the eigenstates of the annihilation operator $\hat{a}$ are all those minimal uncertainty wave packets that have the same spatial width as the oscillator ground state.

Turning now to the *time development* of the state, it is convenient to use the ket notation

$$|\psi(x,t=0)\rangle=|\langle x_0\rangle,\langle p_0\rangle\rangle$$

with $|x,p\rangle$ denoting a minimum uncertainly wave packet having those expectation values of position and momentum.

The time development of the ket, as usual, is given by

$$|\psi(x,t)\rangle=e^{-iHt/\hbar}|\langle x_0\rangle,\langle p_0\rangle\rangle.$$

We shall show that $|\psi(x,t)\rangle$ *remains* an eigenstate of the annihilation operator *for all times t*: it therefore continues to be a minimum uncertainty wave packet!

The key point in establishing this is that the annihilation operator itself has a simple time development in the Heisenberg representation,

$$\hat{a}(t)=e^{iHt/\hbar}\hat{a}e^{-iHt/\hbar}=\hat{a}e^{-i\omega t}.$$

To prove this, consider the matrix elements of $\hat{a}(t)$ between any two eigenstates $|n\rangle$ of the Hamiltonian

$$H|n\rangle=\left(n+\tfrac{1}{2}\right)\hbar\omega|n\rangle$$

so

$$\langle m|\hat{a}(t)|n\rangle=e^{i\left(m+\frac{1}{2}\right)\hbar\omega t/\hbar}\langle m|\hat{a}|n\rangle e^{-i\left(n+\frac{1}{2}\right)\hbar\omega t/\hbar}=\langle n-1|\hat{a}|n\rangle e^{-i\omega t}.$$

Since the only nonzero matrix elements of the annihilation operator $\langle m|\hat{a}|n\rangle$ are for $m=n-1$, and the energy eigenstates form a complete set, this simple time dependence is true as an *operator* equation

$$\hat{a}(t)=e^{iHt/\hbar}\hat{a}e^{-iHt/\hbar}=\hat{a}e^{-i\omega t}.$$

It is now easy to prove that

$$|\psi(x,t)\rangle=e^{-iHt/\hbar}|\langle x_0\rangle,\langle p_0\rangle\rangle$$

is always an eigenstate of $\hat{a}$:

$$\begin{aligned}\hat{a}|\psi(x,t)\rangle&=\hat{a}e^{-iHt/\hbar}|\langle x_0\rangle,\langle p_0\rangle\rangle\\&=e^{-iHt/\hbar}\left(e^{iHt/\hbar}\hat{a}e^{-iHt/\hbar}\right)|\langle x_0\rangle,\langle p_0\rangle\rangle\\&=e^{-iHt/\hbar}e^{-i\omega t}\hat{a}|\langle x_0\rangle,\langle p_0\rangle\rangle\\&=e^{-iHt/\hbar}e^{-i\omega t}\left(m\omega\langle x_0\rangle+i\langle p_0\rangle\right)/\sqrt{2\hbar m\omega}|\langle x_0\rangle,\langle p_0\rangle\rangle\\&=\left(e^{-i\omega t}\left(m\omega\langle x_0\rangle+i\langle p_0\rangle\right)/\sqrt{2\hbar m\omega}\right)|\psi(x,t)\rangle.\end{aligned}$$

Therefore the annihilation operator, which at $t=0$ had the eigenvalue

$$z_0=\left(m\omega\langle x_0\rangle+i\langle p_0\rangle\right)/\sqrt{2\hbar m\omega},$$

corresponding to a minimal wave packet centered at $(m\omega\langle x_0\rangle,\langle p_0\rangle)$ in phase space, evolves in time t to another minimal packet (because it's still an eigenstate of the annihilation operator), and writing

$$\left|\langle x(t)\rangle,\langle p(t)\rangle\right\rangle = e^{-iEt/\hbar}\left|\langle x_0\rangle,\langle p_0\rangle\right\rangle,$$

the new eigenvalue of $\hat{a}$

$$z(t) = \frac{\left(m\omega\langle x(t)\rangle + i\langle p(t)\rangle\right)}{\sqrt{2\hbar m\omega}} = \frac{\left(m\omega\langle x_0\rangle + i\langle p_0\rangle\right)}{\sqrt{2\hbar m\omega}}e^{-i\omega t} = z(0)e^{-i\omega t}.$$

Therefore, the centre of the minimal wave packet in phase space follows the classical path in time. This is made explicit by equating real and imaginary parts:

$$\langle x(t)\rangle = \langle x_0\rangle\cos\omega t + \left(\langle p_0\rangle / m\omega\right)\sin\omega t,$$

$$\langle p(t)\rangle = \langle p_0\rangle\cos\omega t - m\omega\langle x_0\rangle\sin\omega t.$$

So we've found Schrodinger's "best possible" quantum description of a classical oscillator.

COHERENT STATES AND THE CLASSICAL LIMIT

The phase space of a *classical* simple harmonic oscillator is conveniently parameterized using the complex variable $z = (m\omega x + ip)/\sqrt{2\hbar m\omega}$. If the oscillator is initially at z_0, it will describe a circle in the complex plane $z(t) = z_0 e^{-i\omega t}$.

If a *quantum* simple harmonic oscillator is initially in a minimum uncertainty state having the same spatial width as the oscillator ground state, and the state has expectation values of position and momentum denoted by $\left(m\omega\langle x_0\rangle,\langle p_0\rangle\right)$, then it is an eigenstate of the annihilation operator $\hat{a}$ with eigenvalue $z_0 = \left(m\omega\langle x_0\rangle + i\langle p_0\rangle\right)/\sqrt{2\hbar m\omega}$, and we label it $|z_0\rangle$.

$$\hat{a}|z_0\rangle = z_0|z_0\rangle.$$

We have established that as this state develops in time, the wave function continues to be a minimal uncertainty wave packet having constant spatial width, and hence continues to be an eigenstate of the annihilation operator, and that the centre of the wave packet in phase space follows *exactly* the path of the classical oscillator,

$$\hat{a}|z\rangle = z|z\rangle, \quad z = z_0 e^{-i\omega t}.$$

A REMARK ON NOTATION

We have chosen to work with the original position and momentum variables, and the complex parameter expressed as a function of those variables, throughout. We could have used the dimensionless variables introduced in the lecture on the simple harmonic oscillator,

$$\xi = x/b = x\sqrt{m\omega/\hbar}, \qquad \pi = bp/\hbar = p/\sqrt{\hbar m\omega},$$

$\hat{a} = \left(\hat{\xi} + i\hat{\pi}\right)/\sqrt{2}$. This would of course also give $z = (\xi + i\pi)/\sqrt{2}$, a more compact representation, but one more thing to remember.

It's also common to denote the eigenstates of $\hat{a}$ by α, $\hat{a}|\alpha\rangle = \alpha|\alpha\rangle$, very elegant, but we've used z to keep reminding ourselves that this eigenvalue, unlike most of those encountered in quantum mechanics, is a complex number. Finally, some use the dimensionless variables $X = \sqrt{2\hbar/m\omega}\,x$, $P = \sqrt{1/(2m\omega\hbar)}\,p$, differing from ξ, π by a factor of $\sqrt{2}$. The eigenvalue equation for the annihilation operator is very neat in this notation: $\hat{a}|z\rangle = (X + iP)|z\rangle$. We've avoided it, though, because our recommended textbook, Shankar, uses *X*, *P* for the ordinary position and momentum operators.

THE TRANSLATION OPERATOR

It's worth repeating the exercise for the simple case of the oscillator initially at rest a distance $\langle x_0 \rangle$ from the centre. This gives a neat tie-in with the *translation* operator (defined below).

Let us then take the initial state to be

$$\psi(x,0) = Ce^{-m\omega(x-\langle x_0\rangle)^2/2\hbar} = \psi_0\left(x-\langle x_0\rangle\right)$$

where $\psi_0(x)$ is the ground state wave function—so we've moved the packet to the right by $\langle x_0\rangle$.

Now do a Taylor series expansion (taking to be the variable!):

$$\psi_0\left(x-\langle x_0\rangle\right) = \psi_0(x) - \langle x_0\rangle \frac{d}{dx}\psi_0(x) + \frac{\langle x_0\rangle^2}{2!}\frac{d^2}{dx^2}\psi_0(x) - \ldots$$

$$= e^{-\langle x_0\rangle \frac{d}{dx}}\psi_0(x).$$

It's clear from this that the *translation operator* $e^{-\langle x_0\rangle \frac{d}{dx}}$ shifts the wave function a distance $\langle x_0\rangle$ to the right.

Since $\hat{p} = -i\hbar d/dx$, the translation operator can also be written as $e^{-i\langle x_0\rangle\hat{p}/\hbar}$, and from this it can be expressed in terms of $\hat{a}, \hat{a}^\dagger$, since

$$\hat{a} = \frac{1}{\sqrt{2\hbar m\omega}}(m\omega\hat{x} + i\hat{p}), \quad \hat{a}^\dagger = \frac{1}{\sqrt{2\hbar m\omega}}(m\omega\hat{x} - i\hat{p}),$$

($\hat{p}, \hat{x}$ being Hermitian) so

$$\hat{p} = i\sqrt{\frac{\hbar m\omega}{2}}\left(\hat{a}^\dagger - \hat{a}\right).$$

Therefore the displaced ground state wave function can be written

$$\psi_0\left(x-\langle x_0\rangle\right) = e^{-i\langle x_0\rangle\hat{p}/\hbar}\psi_0(x)$$

$$= e^{\langle x_0\rangle\sqrt{m\omega/2\hbar}\left(\hat{a}^\dagger - \hat{a}\right)}\psi_0(x)$$

$$= e^{x_0\left(\hat{a}^\dagger - \hat{a}\right)}\psi_0(x)$$

for *real* $z_0 = \langle x_0 \rangle \sqrt{m\omega / 2\hbar}$, since $\langle p_0 \rangle$ is zero for this initial state (the wave function is real).

In the ket notation, we have established that the minimal uncertainty state centered at x_0, and having zero expectation value for the momentum, is

$$\left|\langle x_0 \rangle, 0\right\rangle = e^{z_0\left(\hat{a}^\dagger - \hat{a}\right)}\left|0,0\right\rangle.$$

But it's not exactly obvious that this is an eigenstate of $\hat{a}$ with eigenvalue z_0! (As it must be.)

It's worth seeing how to prove that just from the properties of the operators—but to do that, we need a couple of theorems concerning exponentials of operators given in the Appendix.

First, if the commutator $[A,B]$ commutes with A and B, then $e^{A+B} = e^A e^B e^{-\frac{1}{2}[A,B]}$. This result simplifies the right hand side of the above equation, for

$$e^{z_0\left(\hat{a}^\dagger - \hat{a}\right)}\left|0,0\right\rangle = e^{z_0\hat{a}^\dagger} e^{-z_0\hat{a}} e^{-z_0^2\left[\hat{a}^\dagger,\hat{a}\right]/2}\left|0,0\right\rangle$$

$$= e^{-z_0^2/2} e^{z_0\hat{a}^\dagger}\left|0,0\right\rangle$$

where we have used $e^{-z_0\hat{a}}\left|0,0\right\rangle = \left|0,0\right\rangle$.

This is simpler, but it's still not obvious that we have an eigenstate of $\hat{a}$: we need the commutator $\left[\hat{a}, e^{z_0\hat{a}^\dagger}\right]$.

The second theorem we need is: if the commutator of two operators $\left[A,B\right] = c$ itself

commutes with A and B, then

$$\left[A, e^{\lambda B}\right] = \lambda c e^{\lambda B}.$$

(This is easily proved by expanding the exponential)

Applying this to our case,

$$\left[\hat{a}, e^{z_0\hat{a}^\dagger}\right] = z_0 e^{z_0\hat{a}^\dagger}.$$

It follows immediately that $e^{-z_0^2/2}e^{z_0\hat{a}^\dagger}\left|0,0\right\rangle$ is indeed an eigenstate of $\hat{a}$ with eigenvalue $z_0 = \langle x_0\rangle\sqrt{m\omega/2\hbar}$. (It must also be correctly normalized because the translation $\left|\langle x_0\rangle,0\right\rangle = e^{z_0(\hat{a}^\dagger-\hat{a})}\left|0,0\right\rangle$ is a unitary operation for real z_0.)

How do we generalize this translation operator to an arbitrary state, with nonzero $\langle x\rangle,\langle p\rangle$? Thinking in terms of the complex parameter space *z*, we need to be able to move in both the *x* and the *p* directions, using both $\hat{p} = -i\hbar d/dx$ and $\hat{x} = i\hbar d/dp$. This is slightly tricky since these operators do not commute, but their commutator is just a number, so (using the theorem proved in the Appendix) this will only affect the overall normalization.

Furthermore, both $\hat{p}$ and $\hat{x}$ are combinations of $\hat{a},\hat{a}^\dagger$, so for the generalization of $e^{-i\langle x_0\rangle\hat{p}/\hbar}$ from real $\langle x_0\rangle$ to complex *z* to be unitary, it must have an *antihermitian* combination of $\hat{a},\hat{a}^\dagger$ in the exponent—a unitary operator has the form $U = e^{iH}$, where *H* is Hermitian, so *iH* is antihermitian..

We are led to the conclusion that

$$\left|\langle p\rangle,\langle x\rangle\right\rangle = e^{(z\hat{a}^\dagger - z^*\hat{a})}\left|0\right\rangle = \left|z\right\rangle,$$

conveniently labeling the coherent state using the complex parameter *z* of its centre in phase space. Since this generalized translation operator is unitary, the new state is automatically correctly normalized.

THE ENERGY EIGENSTATES

The equation above suggests the possibility of representing the displaced state $\left|z\right\rangle$ in the standard energy basis $\left|n\right\rangle$. We can simplify with the same trick used for the spatial displacement case in the last section, that is, the theorem $e^{A+B} = e^A e^B e^{-\frac{1}{2}[A,B]}$ where now $A = z\hat{a}^\dagger$, $B = -z^*\hat{a}$:

$$|z\rangle = e^{z\hat{a}^\dagger - z^*\hat{a}}|0\rangle = e^{-|z|^2/2}e^{z\hat{a}^\dagger}e^{-z^*\hat{a}}|0\rangle = e^{-|z|^2/2}e^{z\hat{a}^\dagger}|0\rangle$$

using $e^{-z^*\hat{a}}|0\rangle = |0\rangle$ since $\hat{a}|0\rangle = 0$.

It is now straightforward to expand the exponential:

$$|z\rangle = e^{-|z|^2/2}e^{z\hat{a}^\dagger}|0\rangle = e^{-|z|^2/2}\left(1 + za^\dagger + \frac{(za^\dagger)^2}{2!} + \ldots\right)|0\rangle$$

and recalling that the normalized energy eigenstates are

$$|n\rangle = \frac{(a^\dagger)^n}{\sqrt{n!}}|0\rangle$$

we find

$$|z\rangle = e^{-|z|^2/2}\left(|0\rangle + z|1\rangle + \frac{z^2}{\sqrt{2!}}|2\rangle + \frac{z^3}{\sqrt{3!}}|3\rangle + \ldots\right).$$

Time Development of an Eigenstate of *a* Using the Energy Basis

Now that we have expressed the eigenstate $|z\rangle$ as a sum over the eigenstates $|n\rangle$ of the Hamiltonian, finding its time development in this representation is straightforward.

Since $|n(t)\rangle = e^{-in\omega t}|n\rangle$,

$$|z(t)\rangle = e^{-|z_0|^2/2}\left(|0\rangle + z_0 e^{-i\omega t}|1\rangle + \frac{z_0^2 e^{-2i\omega t}}{\sqrt{2!}}|2\rangle + \frac{z_0^3 e^{-3i\omega t}}{\sqrt{3!}}|3\rangle + \ldots\right)$$

which can be written

$$|z(t)\rangle = e^{-|z_0|^2/2}e^{z_0 e^{-i\omega t}\hat{a}^\dagger}|0\rangle,$$

equivalent to the result $z(t) = z_0 e^{-i\omega t}$ derived earlier.

Some Properties of the Set of Eigenstates of *a*

In quantum mechanics, any physical variable is represented by a Hermitian operator. The eigenvalues are real, the eigenstates are orthogonal (or can be chosen to be so for degenerate states) and the eigenstates for a complete set, spanning the space, so any vector in the space can be represented in a unique way as a sum over these states.

The operator is not Hermitian. Its eigenvalues are *all the numbers in the complex plane.* The eigenstates belonging to different eigenvalues are never orthogonal, as is immediately obvious on considering the ground state and a displaced ground state. The overlap does of course decrease rapidly for states far away in phase space.

The state overlap can be computed using $|z\rangle = e^{-|z|^2/2} e^{z\hat{a}^\dagger} |0\rangle$:

$$\langle w|z\rangle = \langle 0| e^{w^*\hat{a}} e^{-|w|^2/2} e^{-|z|^2/2} e^{z\hat{a}^\dagger} |0\rangle$$

and we can then switch the operators $e^{-w^*\hat{a}}$, $e^{z\hat{a}^\dagger}$ using the theorem from the Appendix $e^B e^A = e^A e^B e^{-[A,B]}$, then since $\langle 0|\hat{a}^\dagger = \hat{a}|0\rangle = 0$, we're left with

$$\langle w|z\rangle = \langle 0| e^{w^* z} e^{-|w|^2/2} e^{-|z|^2/2} |0\rangle,$$

from which

$$|\langle w|z\rangle|^2 = e^{-|w-z|^2}.$$

Finally, using $|z\rangle = e^{-|z|^2/2}\left(|0\rangle + z|1\rangle + \frac{z^2}{\sqrt{2!}}|2\rangle + \frac{z^3}{\sqrt{3!}}|3\rangle + \dots\right)$, we can construct a unit operator using the $|z\rangle$,

$$I = \iint \frac{dxdy}{\pi} |z\rangle\langle z|$$

where the integral is over the whole complex plane $z = x + iy$ (this x is not, of course, the original position x, recall for the wave function just displaced along the axis $z_0 = \langle x_0\rangle\sqrt{m\omega/2\hbar}$). Therefore, the $|z\rangle$ span the whole space.

SOME EXPONENTIAL OPERATOR ALGEBRA

Suppose that the commutator of two operators A, B

$$[A,B]=c,$$

where c commutes with A and B, usually it's just a number, for instance 1 or $i\hbar$.

Then

$$\left[A,e^{\lambda B}\right]=\left[A,1+\lambda B+\left(\lambda^2/2!\right)B^2+\left(\lambda^3/3!\right)B^3+\dots\right]$$

$$=\lambda c+\left(\lambda^2/2!\right)2Bc+\left(\lambda^3/3!\right)3B^2c+\dots$$

$$=\lambda ce^{\lambda B}.$$

That is to say, the commutator of A with $e^{\lambda B}$ is proportional to itself.

That is reminiscent of the simple harmonic oscillator commutation relation $[H,a^{\dagger}]=\hbar\omega a^{\dagger}$ which led directly to the ladder of eigenvalues of H separated by $\hbar\omega$. Will there be a similar "ladder" of eigenstates of A in general?

Assuming A (which is a general operator) has an eigenstate $|a\rangle$ with eigenvalue a,

$$A|a\rangle=a|a\rangle.$$

Applying $[A,e^{\lambda B}]=\lambda ce^{\lambda B}$ to the eigenstate $|a\rangle$:

$$Ae^{\lambda B}|a\rangle=e^{\lambda B}A|a\rangle+\lambda ce^{\lambda B}|a\rangle=(a+\lambda c)|a\rangle.$$

Therefore, unless it is identically zero, $e^{\lambda B}|a\rangle$ is *also* an eigenstate of A, with eigenvalue $a+\lambda c$. We conclude that instead of a *ladder* of eigenstates, we can apparently generate a whole *continuum* of eigenstates, since λ can be set arbitrarily.

To find more operator identities, premultiply $[A,e^{\lambda B}]=\lambda ce^{\lambda B}$ by $e^{-\lambda B}$ to find:

$$e^{-\lambda B} A e^{\lambda B} = A + \lambda[A, B]$$
$$= A + \lambda c.$$

This identity is *only* true for operators *A*, *B* whose commutator *c* is a number. (Well, *c could* be an operator, provided it still commutes with both *A* and *B*).

Our next task is to establish the following very handy identity, which is also only true if [*A*,*B*] commutes with *A* and *B*:

$$e^{A+B} = e^{A} e^{B} e^{-\frac{1}{2}[A,B]}.$$

The proof (due to Glauber, given in Messiah) is as follows:

Take $f(x) = e^{Ax} e^{Bx}$,

$$df/dx = Ae^{Ax}e^{Bx} + e^{Ax}e^{Bx}B$$
$$= f(x)\left(e^{-Bx}Ae^{Bx} + B\right)$$
$$= f(x)\left(A + x[A, B] + B\right).$$

It is easy to check that the solution to this first-order differential equation equal to one at $x = 0$ is

$$f(x) = e^{x(A+B)} e^{\frac{1}{2}x^2[A,B]}$$

so taking $x = 1$ gives the required identity, $e^{A+B} = e^{A}e^{B}e^{-\frac{1}{2}[A,B]}$.

It also follows that $e^{B}e^{A} = e^{A}e^{B}e^{-[A,B]}$ provided—as always—that [*A*, *B*] commutes with *A* and *B*.

Chapter 16

Path Integrals in Quantum Mechanics

HUYGEN'S PICTURE OF WAVE PROPAGATION

If a point source of light is switched on, the wavefront is an expanding sphere centered at the source. Huygens suggested that this could be understood if at any instant in time each point on the wavefront was regarded as a source of secondary wavelets, and the new wavefront a moment later was to be regarded as built up from the sum of these wavelets. For a light shining continuously, this process just keeps repeating.

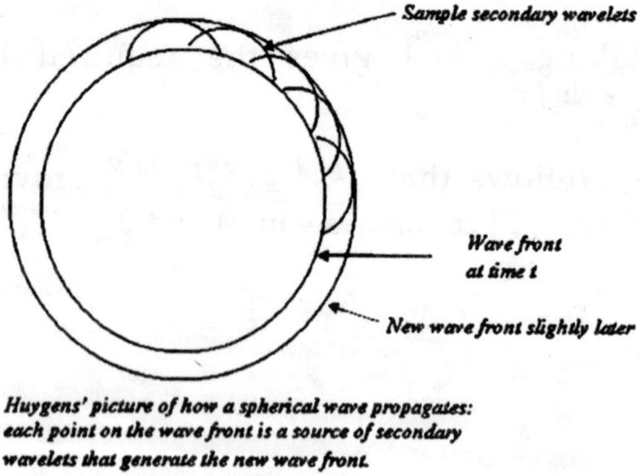

Huygens' picture of how a spherical wave propagates: each point on the wave front is a source of secondary wavelets that generate the new wave front.

What use is this idea? For one thing, it explains refraction—the change in direction of a wavefront on entering a different medium, such as a ray of light going from air into glass.

If the light moves more slowly in the glass, velocity v instead of c, with $v < c$, then Huygen's picture explains Snell's Law, that the ratio of the sines of the angles to the normal of incident and transmitted beams is constant, and in fact is the ratio c/v. This is evident from the diagram below: in the time the wavelet centered at A has propagated to C, that from B has reached D, the ratio of lengths AC/BD being c/v. But the angles in Snell's Law are in fact the angles ABC, BCD, and those right-angled triangles have a common hypotenuse BC, from which the Law follows.

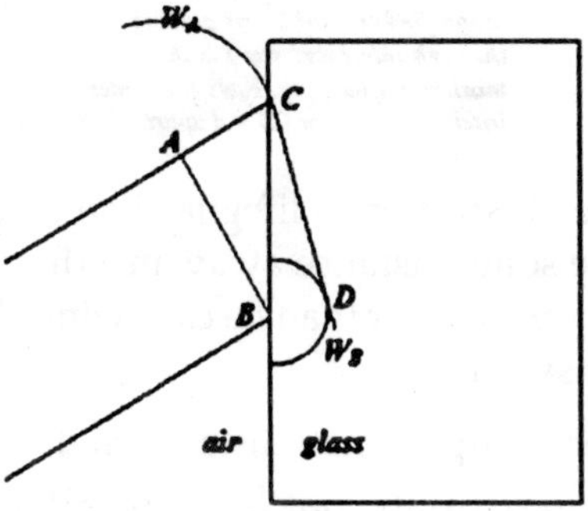

FERMAT'S PRINCIPLE OF LEAST TIME

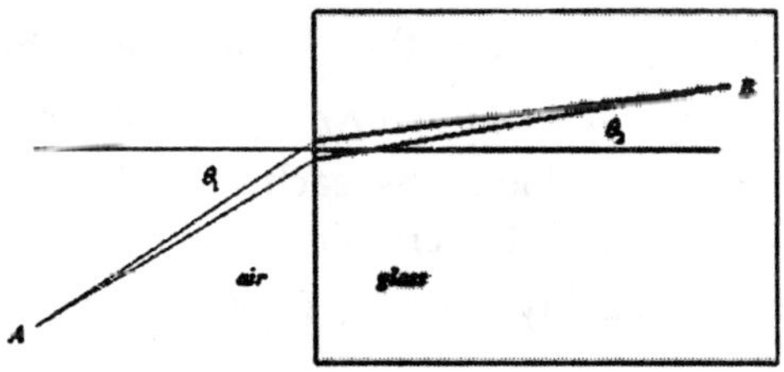

We will now temporarily forget about the wave nature of light, and consider a narrow ray or beam of light shining from point A to point B, where we suppose A to be in air, B in glass. Fermat showed that the path of such a beam is given by the Principle of Least Time: a ray of light going from A to B by any other path would take longer. How can we see that? It's obvious that any deviation from a straight line path in air or in the glass is going to add to the time taken, but what about moving slightly the point at which the beam enters the glass?

Where the air meets the glass, the two rays, separated by a small distance $CD = d$ along that interface, will look parallel:

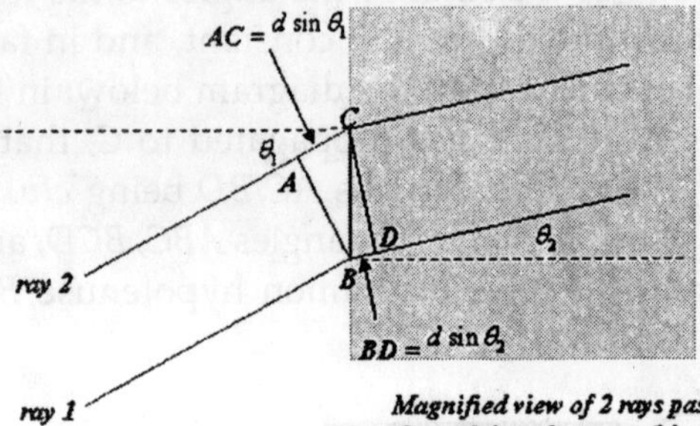

Magnified view of 2 rays passing through interface: ray 1 is the minimum time path. Rays encounter interface distance CB = d apart.

(Feynman gives a nice illustration: a lifeguard on a beach spots a swimmer in trouble some distance away, in a diagonal direction. He can run three times faster than he can swim. What is the quickest path to the swimmer?)

Moving the point of entry up a small distance *d*, the light has to travel an extra $d \sin\theta_1$ in air, but a distance less by $d \sin\theta_2$ in the glass, giving an extra travel time $\Delta t = d \sin\theta_1 / c - d \sin\theta_2 / v$. For the *classical* path, Snell's Law gives $\sin\theta_1 / \sin\theta_2 = n = c / v$, so $\Delta t = 0$ to first order. But if we look at a series of possible paths, each a small distance *d* away from the next at the point of crossing from air into glass, Δt becomes of order *d*/*c* away from the classical path.

Suppose now we imagine that the light actually travels along all these paths with about equal amplitude. (This actually is what Huygen's picture suggests: if we imagine the wavefront to generate secondary wavelets every picosecond, say, we can visualize the paths as zigzags with steps of length 3mm.) What will be the total contribution of all these paths at *B*? Since the times along the paths are different, the signals along the different paths will arrive at B with different phases, and to get the total wave amplitude we must add a series of unit $2D$ vectors, one from each path. (Representing the amplitude and phase of the wave by a complex number for convenience—for a real wave, we can take the real part at the end.)

When we map out these unit $2D$ vectors, we find that in the neighbourhood of the classical path, the phase varies little, but as we go away from it the phase spirals more and more rapidly, so those paths interfere amongst themselves destructively.

This is the explanation of Fermat's Principle—only near the path of least time do paths stay approximately in phase with each other and add constructively. So this classical path rule has an underlying wave-phase explanation. In fact, the central role of phase in this analysis is sometimes emphasized by saying the light beam follows *the path of stationary phase.*

THE PRINCIPLE OF LEAST ACTION

Confining our attention for the moment to the mechanics of a single nonrelativistic particle in a potential, with Lagrangian $L = T - V$, the action S is defined by

$$S = \int_{t_1}^{t_2} L(x, \dot{x})dt.$$

Newton's Laws of Motion can be shown to be equivalent to the statement that a particle moving in the potential from A at t_1 to B at t_2 travels along the path that minimizes the action. This is called the *Principle of Least Action*: for example, the parabolic path followed by a ball thrown through the air minimizes the integral along the path of the action T-V where T is the ball's kinetic energy, V its gravitational potential energy (neglecting air resistance, of course). Note here that the initial and final times are fixed, so since we'll be summing over paths with different lengths, necessarily the particles speed will be different along the different paths. In other words, it will have different energies along the different paths.

With the advent of quantum mechanics, and the realization that any particle, including a thrown ball, has wave like properties, the rather mysterious Principle of Least Action looks a lot like Fermat's Principle of Least Time. Recall that Fermat's Principle works because the total phase along a path is the integrated time elapsed along the path, and for a path where that integral is stationary for small path variations, neighbouring paths add constructively, and no other sets of paths do. If the Principle of Least Action has a similar explanation, then the wave amplitude for a particle going along a path from A to B must have a phase equal to some constant times the action along that path. If this is the case, then the observed path followed will be just that of least action, for only near that path will the amplitudes add constructively, just as in Fermat's analysis of light rays.

GOING FROM CLASSICAL MECHANICS TO QUANTUM MECHANICS

Of course, if we write a phase factor for a path e^{icS} where S is the action for the path and c is some constant, c must necessarily have the dimensions of inverse action. Fortunately, there is a natural candidate for the constant c. The wave nature of matter arises from quantum mechanics, and the fundamental constant of quantum mechanics, Planck's constant, is in fact a unit of action. It turns out that the appropriate path phase factor is $e^{iS/\hbar}$.

That the phase factor is $e^{iS/\hbar}$, rather than $e^{iS/\hbar}$, say, can be established by considering the double slit experiment for electrons

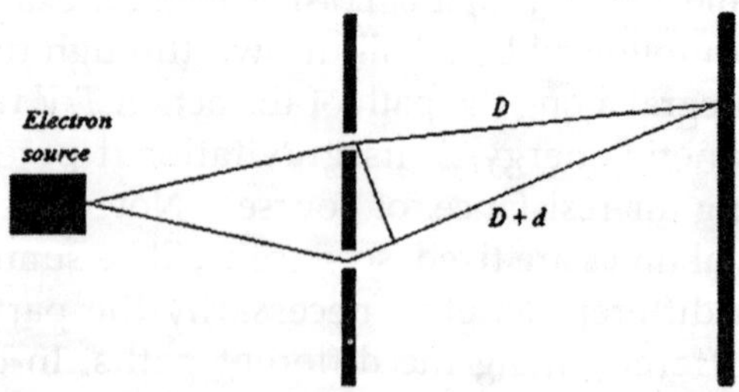

Suppose electrons from the top slit, Path I, go a distance D to the detector, those from the bottom slit, Path II, go $D + d$, with $d \ll D$. Then if the electrons have wavelength λ we know the phase difference at the detector is $2\pi d / \lambda$. To see this from our formula for summing over paths, on Path I the action $S = Et = ½mv_1^2t$, and $v_1 = D/t$, so

$$S_1 = ½mD^2/t.$$

For Path II, we must take $v_2 = (D + d)/t$. Keeping only terms of leading order in d/D, the action difference between the two paths

$$S_2 - S_1 = mDd/t$$

So the phase difference

$$\frac{S_2 - S_1}{\hbar} = \frac{mvd}{\hbar} = \frac{2\pi pd}{h} = \frac{2\pi d}{\lambda}.$$

This is the known correct result, and this fixes the constant multiplying the action/h in the expression for the path phase.

In quantum mechanics, such as the motion of an electron in an atom, we know that the particle does not follow a well-defined path, in contrast to classical mechanics. Where does the crossover to a well-defined path take place? Taking the simplest possible case of a free particle (no potential) of mass m moving at speed v, the action along a straight line path taking time t from A to B is $½mv^2t$. If this action is of order Planck's constant h, then the phase factor will not oscillate violently on moving to different paths, and a range of paths will contribute. In other words, quantum rather than classical behaviour dominates when $½mv^2t$ is of order h. But vt is the path length L, and mv/h is the wavelength , so we conclude that we must use quantum mechanics when the wavelength h/p is significant compared with the path length. Interference sets in when the *difference* in path actions is of order h, so in the atomic regime many paths must be included.

Feynman (in Feynman and Hibbs) gives a nice picture to help think about summing over paths. He begins with the double slit experiment for an electron. We suppose the electron is emitted from some source A on the left, and we look for it at a point B on a screen to the right. In the middle is a thin opaque barrier with the familiar two slits. Evidently, to find the amplitude for the electron to reach B we sum over two paths. Now suppose we add *another* two-slit barrier. We have to sum over four paths. Now add another. Next, replace the two slits in each barrier by several slits. We must sum over a multitude of paths! Finally, increase the number of barriers to some large number N, and at the same time increase the number of slits to the point that there are no barriers left. We are left with a sum over all possible paths through space from A to B, multiplying each path by the appropriate action phase factor.

In fact, the sum over paths is even more daunting than this picture suggests. All the paths going through these many slitted barriers are progressing in a forward direction, from A towards B. Actually, if we're summing over all paths, we should be including the possibility of paths zigzagging backwards and forwards as well, eventually arriving at B. We shall soon see how to deal systematically with all possible paths.

THE FREE ELECTRON PROPAGATOR

As a warm up exercise, we consider an electron confined to one dimension, with no potential present, moving from x′ at time 0 to x at time t. (This is speaking loosely—we mean, as explained previously, that the initial state of the electron is a normalizable state, such as a Gaussian, concentrated closely at x′.. The propagator then represents the probability amplitude, that is, the wave function, at point x after the given time t.) The propagator is given by

$$|\psi(x,t)\rangle = U(t)|\psi(x,t=0)\rangle,$$

or, in Schrodinger wave function notation,

$$\psi(x,t) = \int U(x,t;x',t=0)\psi(x',t=0)dx'.$$

It is clear that for this to make sense, as

$$t \to 0, \quad U(x,t;x',0) \to \delta(x-x').$$

In the lecture on propagators, we found

$$\langle x|U(t,0)|x'\rangle = \int_{-\infty}^{\infty} e^{-i\hbar k^2 t/2m} \frac{dk}{2\pi} \langle x|k\rangle\langle k|x'\rangle$$

$$= \int_{-\infty}^{\infty} e^{-i\hbar k^2 t/2m} \frac{dk}{2\pi} e^{-ik(x-x')}$$

$$= \sqrt{\frac{m}{2\pi\hbar i t}} e^{im(x-x')^2/2\hbar t}.$$

Now let us think about the sum over paths. Let us *assume* that the classical path dominates, and that only paths in its neighbourhood contribute, and all the other paths do is multiply the effect of the single classical path by some function of time. (This arbitrary seeming assumption depends heavily on knowing the answer in advance. The classical path, of course, corresponds to motion from x' to x at a constant speed $v = (x - x')/t$. The action along this path is therefore Et, where E is the classical energy $\frac{1}{2} mv^2$, giving

$$U(x,t;x',0) = A' e^{im(x-x')^2/2\hbar t}.$$

This gives the correct exponential term! The prefactor A¢ can be determined from the requirement that as t goes to zero, U must approach a δ-function. This gives the correct prefactor, identical to the one found previously.

However, we have been lucky—in more interesting situations, the classical path doesn't give all the information, and we really must address the issue of integrating over all paths.

Proving that the Sum-Over-Paths Definition of the Propagator is Equivalent to the Sum-Over-Eigenfunctions Definition

The first step is to construct a practical method of summing over paths. Let us begin with a particle in one dimension going from x' at time t' to x at time t. The paths can be enumerated in a crude way, reminiscent of Riemann integration: divide the time interval t' to t into N equal intervals each of duration ε, so $t_0 = t', t_1 = t_0 + \varepsilon, t_2 = t_0 + 2\varepsilon, \ldots, t_N = t.$

Next, define a particular path from x to x' by specifying the position of the particle at each of the intermediate times, that is to say, it is at x_1 at time t_1, x_2 at time t_2 and so on. Then, simplify the path by putting in straight line bits connecting x_0 to x_1, x_1 to x_2, etc. The justification is that in the limit of going to zero, taken at the end, this becomes a true representation of the path.

The next step is to sum over all possible paths with a factor $e^{iS/\hbar}$ for each one. The sum is accomplished by integrating over all possible values of the intermediate positions x_1, x_2, ... x_{N-1} and then taking N to infinity.

The action on the zigzag path is

$$S = \int_{t'}^{t} dt(\tfrac{1}{2}m\dot{x}^2 - V(x)) \rightarrow \sum_i \left[\frac{m(x_{i+1} - x_i)^2}{2\varepsilon} - \varepsilon V\left(\frac{x_{i+1} + x_i}{2}\right)\right]$$

We define the "integral over paths" written $\int D[x(t)]$ by

$$\lim_{\substack{\varepsilon \to 0 \\ N \to \infty}} \frac{1}{B(\varepsilon)} \int_{-\infty}^{\infty}\int \ldots \int \frac{dx_1}{B(\varepsilon)} \ldots \frac{dx_{N-1}}{B(\varepsilon)}$$

where we haven't yet figured out what the overall weighting factor $B(\varepsilon)$ is going to be. (It is standard convention to have that extra $B(\varepsilon)$ outside.)

To summarize: the propagator $U(x,t;x',t')$ is the contribution to the wave function at x at time t from that at x' at the earlier time t'

Consequently, $U(x,t;x',t')$ regarded as a function of x, t is, in fact, nothing but the Schrodinger wave function $\psi(x,t)$, and therefore must *satisfy Schrodinger's equation.*

$$i\hbar\frac{\partial}{\partial t}U(x,t;x',t') = \left(-\frac{\hbar^2}{2m}\frac{\partial^2}{\partial x^2}+V(x)\right)U(x,t;x',t').$$

We shall now show that defining $U(x,t;x',t')$ as a sum over paths, it does in fact satisfy Schrodinger's equation, and furthermore goes to a δ-function as time goes to zero.

$$U(x,t;x',t') = \int D[x(t)]e^{iS[x(t)]/\hbar}$$

$$= \lim_{\substack{\varepsilon\to 0\\ N\to\infty}} \frac{1}{B(\varepsilon)}\int_{-\infty}^{\infty}\int\ldots\int\frac{dx_1}{B(\varepsilon)}\cdots\frac{dx_{N-1}}{B(\varepsilon)}e^{iS(x_1,\ldots,x_{N-1})/\hbar}.$$

We shall establish this equivalence by proving that it satisfies the same differential equation. It clearly has the same initial value—as t' and t coincide, it goes to $\delta(x-x')$ in both representations.

To differentiate $U(x,t;x',t')$ with respect to *t*, we isolate the integral over the last path variable, x_{N-1}:

$$U(x,t;x',t') = \int\frac{dx_{N-1}}{B(\varepsilon)}e^{\left[\frac{im(x-x_{N-1})^2}{2\hbar\varepsilon}-\frac{i}{\hbar}\varepsilon V\left(\frac{x+x_{N-1}}{2}\right)\right]}U(x_{N-1},t-\varepsilon,x',t')$$

Now in the limit ε going to zero, almost all the contribution to this integral must come from close to the point of stationary phase, that is, $x_{N-1} = x$. In that limit, we can take $U(x_{N-1},t-\varepsilon,x',t')$ to be a slowly varying function of x_{N-1}, and replace it by the leading terms in a Taylor expansion about *x*, so

$$U(x,t,x',t') =$$

$$\int\frac{dx_{N-1}}{B(\varepsilon)}e^{\frac{im(x-x_{N-1})^2}{2\hbar\varepsilon}}\left(1-\frac{i}{\hbar}\varepsilon V\left(\frac{x+x_{N-1}}{2}\right)\right)\left(U(x,t-\varepsilon)+(x_{N-1}-x)\frac{\partial U}{\partial x}+\right.$$

$$\left.\frac{(x_{N-1}-x)^2}{2}\frac{\partial^2 U}{\partial x^2}\right)$$

The x_{N-1} dependence in the potential *V* can be neglected in leading order—that leaves standard Gaussian integrals, and

$$U(x,t;x't') = \frac{1}{B(\varepsilon)}\sqrt{\frac{2\pi\hbar\varepsilon}{-im}}\left(1-\frac{i\varepsilon}{\hbar}V(x)+\frac{i\varepsilon\hbar}{2m}\frac{\partial^2}{\partial x^2}\right)U(x,t-\varepsilon;x't').$$

Taking the limit of going to zero fixes our unknown normalizing factor,

$$B(\varepsilon) = \sqrt{\frac{2\pi\hbar\varepsilon}{-im}}$$

giving

$$i\hbar\frac{\partial}{\partial t}U(x,t;x',t') = \left(-\frac{\hbar^2}{2m}\frac{\partial^2}{\partial x^2}+V(x)\right)U(x,t;x',t'),$$

thus establishing that the propagator derived from the sum over paths obeys Schrodinger's equation, and consequently gives the same physics as the conventional approach.

THE PATH INTEGRAL FOR THE FREE PARTICLE

The required correspondence to the Schrodinger equation result fixes the unknown normalizing factor, as we've just established. This means we are now in a position to evaluate the sum over paths explicitly, at least in the free particle case, and confirm the somewhat handwaving result given above.

The sum over paths is

$$U(x,t;x',t') = \int D[x(t)]e^{iS[x(t)]/\hbar} = \lim_{\substack{\varepsilon\to 0\\ N\to\infty}}\frac{1}{B(\varepsilon)}\int_{-\infty}^{\infty}\int\ldots\int\frac{dx_1}{B(\varepsilon)}\ldots\frac{dx_{N-1}}{B(\varepsilon)}e^{i\sum_i\frac{m(x_{i+1}-x_i)^2}{2\hbar\varepsilon}}.$$

Let us consider the sum for small but finite ε. In particular, we'll divide up the interval first into halves, then quarters, and so on, into 2^n small intervals. The reason for this choice will become clear.

Now, we'll integrate over half the paths: those for i odd, leaving the even x_i values fixed for the moment. The integrals are of the form

$$U(x,t;x',t') = \int D[x(t)]e^{iS[x(t)]/\hbar} = \lim_{\substack{\varepsilon\to 0\\ N\to\infty}} \frac{1}{B(\varepsilon)} \int_{-\infty}^{\infty}\int \ldots \int \frac{dx_1}{B(\varepsilon)}$$

$$\ldots \frac{dx_{N-1}}{B(\varepsilon)} e^{i\sum_i \frac{m(x_{i+1}-x_i)^2}{2\hbar\varepsilon}}.$$

using the standard result $\int_{-\infty}^{\infty} dx e^{-ax^2+bx} = \sqrt{\frac{\pi}{a}} e^{b^2/4a}.$

Now put in the value $a = m/\hbar\varepsilon$: the factor $\sqrt{\frac{\pi}{-ia}} = \sqrt{\frac{\pi\hbar\varepsilon}{-im}}$ cancels the normalization factor $B(\varepsilon) = \sqrt{\frac{2\pi\hbar\varepsilon}{-im}}$ except for the factor of 2 inside the square root. But we need that factor of 2, because we're left with an integral—over the remaining even numbered paths—exactly like the one before except that the time interval has doubled, both in the normalization factor and in the exponent, $\varepsilon \to 2\varepsilon$.

So we're back where we started. We can now repeat the process, halving the number of paths again, then again, until finally we have the same expression but with only the fixed endpoints appearing.

Chapter 17

Angular Momentum

TRANSLATION AND ROTATION OPERATORS

As a warm up to analyzing how a wave function transforms under rotation, we review the effect of *linear translation* on a single particle wave function $\psi(x)$. We have already seen an example of this: the coherent states of a simple harmonic oscillator discussed earlier were (at $t = 0$) identical to the ground state except that they were centered at some point displaced from the origin. In fact, the operator creating such a state from the ground state is a translation operator.

The *translation operator* $T(a)$ is *defined* at that operator which when it acts on a wave function ket $|\psi(x)\rangle$ gives the ket corresponding to that wave function moved over by a, that is,

$$T(a)|\psi(x)\rangle = |\psi(x-a)\rangle,$$

so, for example, if $\psi(x)$ is a wave function centered at the origin, $T(a)$ moves it to be centered at the point a.

We have written the wave function as a ket here to emphasize the parallels between this operation and some later ones, but it is simpler at this point to just work with the wave function as a function, so we will drop the ket bracket for now. The form of $T(a)$ as an operator on a function is made evident by rewriting the Taylor series in operator form:

$$\psi(x-a) = \psi(x) - a\frac{d}{dx}\psi(x) + \frac{a^2}{2!}\frac{d^2}{dx^2}\psi(x) - \ldots$$

$$= e^{-a\frac{d}{dx}}\psi(x)$$

$$= T(a)\psi(x).$$

Now for the quantum connection: the differential operator appearing in the exponential is in quantum mechanics proportional to the momentum operator ($p = -i\hbar d/dx$) so the translation operator

$$T(a) = e^{-iap/\hbar}.$$

An important special case is that of an infinitesimal translation,

$$T(\varepsilon) = e^{-i\varepsilon p/\hbar} = 1 - i\varepsilon p/\hbar$$

The momentum p is said to be the *generator* of the translation.

$T(\varepsilon)|x\rangle = |x+\varepsilon\rangle$. Here $|x\rangle$ denotes a delta-function type wave function centered at x. It might be better if he had written $T(\varepsilon)|x_0\rangle = |x_0+\varepsilon\rangle$, then we would see right away that this translates into the wave function transformation $T(\varepsilon)\delta(x-x_0) = \delta(x-x_0-\varepsilon)$, the sign of ε now obviously consistent with our usage above.)

It is important to be clear about whether the *system* is being translated by *a*, as we have done above or whether, alternately, the *coordinate axes* are being translated by *a*, that latter would result in the *opposite* change in the wave function. Translating the coordinate axes, along with the apparatus and any external fields by *-a* relative to the wave function would of course give the same physics as translating the wave function by *+a*. In fact, these two equivalent operations are analogous to the time development of a wave function being described either by a Schrodinger picture, in which the bras and kets change in time, but not the operators, and the Heisenberg picture in which the operators develop but the bras and kets do not change. To pursue this analogy a little further, in the "Heisenberg" case

$$x \to T^{-1}(\varepsilon)xT(\varepsilon) = e^{i\varepsilon p/\hbar}xe^{-i\varepsilon p/\hbar} = x + i\varepsilon[p,x]/\hbar = x + \varepsilon$$

and p is unchanged since it commutes with the operator. So there are two possible ways to deal with translations: transform the bras and kets, *or* transform the operators. We shall almost always leave the operators alone, and transform the bras and kets.

We have established that *the momentum operator is the generator of spatial translations* (the generalization to three dimensions is trivial). We know from earlier work that the Hamiltonian is the generator of *time* translations, by which we mean

$$\psi(t+a) = e^{-iHa/\hbar}\psi(t).$$

It is tempting to conclude that the *angular momentum* must be the operator generating *rotations* of the system, and, in fact, it is easy to check that this is correct. Let us consider an infinitesimal rotation $\delta\vec{\theta}$ about some axis through the origin (the infinitesimal vector being in the direction of the axis). A wavefunction $\psi(\vec{r})$ initially localized at $\vec{r}_0$ will shift to be localized at $\vec{r}_0 + \delta\vec{r}_0$, where $\delta\vec{r}_0 = \delta\vec{\theta} \times \vec{r}_0$. So, how does a wave function transform under this small rotation? Just as for the translation case, $\psi(\vec{r}) \to \psi(\vec{r} - \delta\vec{r})$. If you don't understand the minus sign, reread the discussion on translations and the sign of ε.

Thus

$$\psi(\vec{r}) \to \psi(\vec{r}) - \frac{i}{\hbar}\delta\vec{r}.\hat{\vec{p}}\psi(\vec{r})$$

to first order in the infinitesimal quantity, so the rotation operator

$$R(\delta\vec{\theta})\psi(\vec{r}) = \left(1 - \frac{i}{\hbar}\delta\vec{\theta} \times \vec{r}.\hat{\vec{p}}\right)\psi(\vec{r})$$

$$= \left(1 - \frac{i}{\hbar}\delta\vec{\theta}.\vec{r}\times\hat{\vec{p}}\right)\psi(\vec{r})$$

$$= \left(1 - \frac{i}{\hbar}\delta\vec{\theta}.\hat{\vec{L}}\right)\psi(\vec{r}).$$

If we write this as

$$R\left(\delta\vec{\theta}\right)\psi(\vec{r}) = e^{-\frac{i}{\hbar}\delta\vec{\theta}.\hat{\vec{L}}}\psi(\vec{r})$$

it is clear that a finite rotation is given by multiplying together a large number of these operators, which just amounts to replacing $\delta\vec{\theta}$ by $\vec{\theta}$ in the exponential. Another way of going from the infinitesimal rotation to a full rotation is to use the identity

$$\lim_{N\to\infty}\left(1 + \frac{A\theta}{N}\right)^N = e^{A\theta}$$

which is clearly valid even if *A* is an operator.

We have therefore established that the orbital angular momentum operator $\hat{\vec{L}}$ is the generator of spatial rotations, by which we mean that if we rotate our apparatus, and the wave function with it, the appropriately transformed wave function is generated by the action of $R\left(\vec{\theta}\right)$ on the original wave function. It is perhaps worth giving an explicit example: suppose we rotate the system, and therefore the wave function, through an infinitesimal angle $\delta\theta_z$ about the z-axis. Denote the rotated wave function by $\psi_{rot}(x,y)$. Then

$$\psi_{rot}(x,y) = \left(1 - \frac{i}{\hbar}(\delta\theta_z)L_z\right)\psi(x,y)$$

$$= \left(1 - \frac{i}{\hbar}(\delta\theta_z)\left(-i\hbar\left(x\frac{d}{dy} - y\frac{d}{dx}\right)\right)\right)\psi(x,y)$$

$$= \left(1-(\delta\theta_z)\left(x\frac{d}{dy}-y\frac{d}{dx}\right)\right)\psi(x,y)$$

$$= \psi\left(x+(\delta\theta_z)y, y-(\delta\theta_z)x\right).$$

That is to say, the value of the new wave function at (x,y) is the value of the old wave function at the point which was rotated into (x,y).

QUANTUM GENERALIZATION OF THE ROTATION OPERATOR

However, it has long been known that in quantum mechanics, orbital angular momentum is *not* the whole story. Particles like the electron are found experimentally to have an internal angular momentum, called spin. In contrast to the spin of an ordinary macroscopic object like a spinning top, the electron's spin is *not* just the sum of orbital angular momenta of internal parts, and any attempt to understand it in that way leads to contradictions.

To take account of this new kind of angular momentum, we generalize the orbital angular momentum $\hat{\vec{L}}$ to an operator $\hat{\vec{J}}$ which is *defined* as the generator of rotations on *any* wave function, including possible spin components, so

$$R(\delta\vec{\theta})\psi(\vec{r}) = e^{-\frac{i}{\hbar}\delta\vec{\theta}\cdot\hat{\vec{J}}}\psi(\vec{r}).$$

This is of course identical to the equation we found for L, but there we derived if from the quantum angular momentum operator including the momentum components written as differentials. But up to this point $\psi(\vec{r})$ has just been a complex valued function of position. From now on, the wave function at a point can have several components, so it is in some vector space, and the rotation operator will operate in this space as well as being a differential operator with respect to position. For example, the wave function could be a vector at each point, so rotation of the system could rotate this vector as well as moving it to a different $\vec{r}$.

To summarize: $\psi(\vec{r})$ is in general an n-component function at each point in space, $R(\delta\vec{\theta})$ is an $n \times n$ matrix in the component space, and the above equation is the *definition* of J. Starting from this definition, we will find J's properties.

The first point to make is that in contrast to translations, rotations do not commute even for a classical system. Rotating a book through $\pi/2$ first about the z-axis then about the x-axis leaves it in a different orientation from that obtained by rotating from the same starting position first $\pi/2$ about the x-axis then $\pi/2$ about the z-axis. Even small rotations do not commute, although the commutator is second order. Since the R-operators are representations of rotations, they will reflect this commutativity structure, and we can see just how they do that by considering ordinary classical rotations of a real vector in three-dimensional space.

The matrices rotating a vector by θ about the x, y and z axes are

$$R_x(\theta)=\begin{pmatrix}1 & 0 & 0\\ 0 & \cos\theta & -\sin\theta\\ 0 & \sin\theta & \cos\theta\end{pmatrix},\quad R_y(\theta)=\begin{pmatrix}\cos\theta & 0 & \sin\theta\\ 0 & 1 & 0\\ -\sin\theta & 0 & \cos\theta\end{pmatrix},$$

$$R_z(\theta)=\begin{pmatrix}\cos\theta & -\sin\theta & 0\\ \sin\theta & \cos\theta & 0\\ 0 & 0 & 1\end{pmatrix}.$$

In the limit of rotations about infinitesimal angles (ignoring higher order terms),

$$R_x(\varepsilon)=1+\varepsilon\begin{pmatrix}0 & 0 & 0\\ 0 & 0 & -1\\ 0 & 1 & 0\end{pmatrix},\quad R_y(\varepsilon)=1+\varepsilon\begin{pmatrix}0 & 0 & 1\\ 0 & 0 & 0\\ -1 & 0 & 0\end{pmatrix},$$

$$R_z(\varepsilon)=1+\varepsilon\begin{pmatrix}0 & -1 & 0\\ 1 & 0 & 0\\ 0 & 0 & 0\end{pmatrix}.$$

It is easy to check that

$$\left[R_x(\varepsilon), R_y(\varepsilon)\right] = \varepsilon^2 \begin{pmatrix} 0 & -1 & 0 \\ 1 & 0 & 0 \\ 0 & 0 & 0 \end{pmatrix},$$

which can also be written

$$R_x(\varepsilon) R_y(\varepsilon) = R_z(\varepsilon^2) R_y(\varepsilon) R_x(\varepsilon).$$

The rotation operators on quantum mechanical kets must, like all rotations, follow this same pattern, that is, we must have

$$\left((1-\frac{i}{\hbar}\varepsilon J_x)(1-\frac{i}{\hbar}\varepsilon J_y)-(1-\frac{i}{\hbar}\varepsilon^2 J_z)(1-\frac{i}{\hbar}\varepsilon J_y)(1-\frac{i}{\hbar}\varepsilon J_x)\right)|\psi\rangle = 0$$

where we have used the definition of the infinitesimal rotation operator on kets, $R(\delta\vec{\theta})\psi(\vec{r}) = e^{-\frac{i}{\hbar}\delta\vec{\theta}\cdot\vec{J}}\psi(\vec{r})$. The zeroth and first-order terms in e all cancel, the second-order term gives $\left[J_x, J_y\right] = i\hbar J_z$. The general statement is:

$$[j_i, j_i] = ih \in_{ijk} J_k$$

This is one of the most important formulas in quantum mechanics.

CONSEQUENCES OF THE COMMUTATION RELATIONS

The commutation formula $\left[J_i, J_j\right] = i\hbar\varepsilon_{ijk}J_k$, which is, after all, a straightforward extension of the result for ordinary classical rotations, has surprisingly far-reaching consequences: it leads directly to the directional quantization of spin and angular momentum observed in atoms subject to a magnetic field.

It is by now very clear that in quantum mechanical systems such as atoms the total angular momentum, and also the component of angular momentum in a given direction, can only take certain values. Let us try to construct a basis set of angular momentum states for a given system: a complete set of kets corresponding to all allowed values of the angular momentum. Now, angular momentum is a *vector* quantity: it has magnitude and direction. Let's begin with the magnitude, the natural parameter is the length squared:

$$J^2 = J_x^2 + J_y^2 + J_z^2.$$

Now we must specify direction—but here we run into a problem .J_x, J_y, and J_z are all mutually non-commuting, so we cannot construct a set of common eigenkets of any two of them, which we would need for a precise specification of direction. They *do* all commute with J^2, since it is spherically symmetric and therefore cannot be affected by any rotation (and, it's easy to check this commutation explicitly).

The bottom line, then, is that in attempting to construct eigenkets describing the different possible angular momentum states of a quantum system, the best we can do is to find the common eigenkets of J^2 and *one* direction, say J_z. The commutation relations do not allow us to be more precise about direction, analogous to the Uncertainty Principle for position and momentum, which also comes from noncommutativity of the relevant operators.

We conclude that the appropriate angular momentum basis is the set of common eigenkets of the commuting Hermitian matrices J^2, J_z:

$$J^2 |a,b\rangle = a |a,b\rangle$$

$$J_z |a,b\rangle = b |a,b\rangle.$$

Our next task is to find the allowed values of *a* and *b*.

LADDER OPERATORS

The sets of allowed eigenvalues *a,b* can be found using the "ladder operator" trick previously discussed for the simple harmonic oscillator.

$$J_\pm = J_x \pm iJ_y$$

are closely analogous to the simple harmonic oscillator raising and lowering operators $a^\dagger$ and a.

J_+ and J_- have commutation relations with J_z:

$$[J_z, J_\pm] = \pm\hbar J_\pm$$

and they of course *commute* with J^2, as do J_z, J_x and J_y.

Therefore, $J_\pm$ operating on $|a,b\rangle$ cannot affect the value of a. But they *do* change the value of b:

$$\begin{aligned} J_z J_\pm |a,b\rangle &= [J_z, J_\pm]|a,b\rangle + J_\pm J_z |a,b\rangle \\ &= \pm\hbar J_\pm |a,b\rangle + bJ_\pm |a,b\rangle \\ &= (b \pm \hbar) J_\pm |a,b\rangle \end{aligned}$$

so if $|a,b\rangle$ is an eigenket of J_z with eigenvalue b, $J_\pm |a,b\rangle$ is either zero or an eigenket of J_z with eigenvalue $b \pm \hbar$, that is, $J_\pm |a,b\rangle = C_\pm |a, b \pm \hbar\rangle$ where $C_\pm(a,b)$ is a normalization constant, taking the initial $|a,b\rangle$ to be normalized. Just as with the simple harmonic oscillator, we have to find these normalization constants in order to compute matrix elements. All the physics is in the matrix elements.

The squared norm of $J_\pm |a,b\rangle$

$$\left\| J_\pm |a,b\rangle \right\|^2 = \langle a,b | J_\pm^\dagger J_\pm | a,b\rangle = \langle a,b | J_\mp J_\pm | a,b\rangle$$

and

$$\begin{aligned} J_\mp J_\pm &= (J_x \mp iJ_y)(J_x \pm iJ_y) = J_x^2 + J_y^2 \pm i[J_x, J_y] \\ &= J^2 - J_z^2 \mp \hbar J_z \end{aligned}$$

from which

$$\left\| J_\pm |a,b\rangle \right\|^2 = \langle a,b | J^2 - J_z^2 \mp \hbar J_z | a,b\rangle = a - b^2 \mp \hbar b,$$

recalling that $\langle a,b|a,b\rangle=1$.

Now *a*, being the eigenvalue of a sum of squares of Hermitian operators, is necessarily nonnegative, and *b* is real. Hence for a given *a*, *b* is *bounded*: there must be a b_{max} and a (negative or zero) b_{min}. But this must mean that

$$\left\|J_+|a,b_{max}\rangle\right\|^2=a-b_{max}^2-\hbar b_{max}=0,$$

$$\left\|J_-|a,b_{min}\rangle\right\|^2=a-b_{min}^2+\hbar b_{min}=0.$$

Note that for a given *a*, b_{max} and b_{min} are determined uniquely—there cannot be two kets with the same *a* but different *b* annihilated by J_+. It also follows immediately that $a=b_{max}(b_{max}+\hbar)$ and $b_{min}=-b_{max}$. Furthermore, we know that if we keep operating on $|a,b_{min}\rangle$ with J_+, we generate a sequence of kets with J_z eigenvalues $b_{min}+\hbar,\ b_{min}+2\hbar,\ b_{min}+3\hbar,\ldots$. This series must terminate, and the only possible way for that to happen is for b_{max} to be equal to $b_{min}+n\hbar$ with *n* an integer, from which it follows that b_{max} is either an integer or half an odd integer times $\hbar$.

At this point, we switch to the standard notation. We have established that the eigenvalues of J_z form a finite ladder, spacing $\hbar$. We write them as $m\hbar$, and *j* is used to denote the maximum value of *m*, so the eigenvalue of J^2, $a=j(j+1)\hbar^2$. Both *j* and *m* will be integers or half odd integers, but the *spacing* of the ladder of *m* values is always unity. Although we have been writing $|a,b\rangle$ with $a=j(j+1)\hbar^2$, $b=m\hbar$ we shall henceforth follow convention and write $|j,m\rangle$.

The operators $\vec{J}^2, J_z$ have a common set of orthonormal eigenkets $|j,m\rangle$,

$$\vec{J}^2 |j,m\rangle = j(j+1)\hbar^2 |j,m\rangle$$
$$J_z |j,m\rangle = m\hbar |j,m\rangle$$
$$\langle j,m | j,m\rangle = 1$$

where j, m are integers or half integers. The allowed quantum numbers m form a ladder with step spacing unity, the maximum value of m is j, the minimum value is $-j$.

NORMALIZING J_+ AND J_-

It is now straightforward to compute the normalization factors needed to find matrix elements:

$$\left\| J_\pm |j,m\rangle \right\|^2 = \langle j,m | J^2 - J_z^2 \mp \hbar J_z | j,m\rangle =$$
$$\left(j(j+1)\hbar^2 - m(m\pm 1)\hbar^2 \right) \langle j,m | j,m\rangle,$$

and $\langle j,m | j,m\rangle = 1$, so

$$J_+ |j,m\rangle = \sqrt{j(j+1) - m(m+1)}\,\hbar\, |j,m+1\rangle$$
$$J_- |j,m\rangle = \sqrt{j(j+1) - m(m-1)}\,\hbar\, |j,m-1\rangle.$$

With these formulas, and the base set of normalized eigenkets $|j,m\rangle$, we are in a position to construct explicit matrix representations of the angular momentum algebra for any integer or half integer value of angular momentum j.

HISTORICAL NOTE:

The use of m to denote the component of angular momentum in one direction came about because a Bohr-type electron in orbit is a current loop, with a magnetic moment parallel to its angular momentum, so the m measured the component of magnetic moment in a chosen direction, usually along an external magnetic field, and m is often called the magnetic quantum number.

Chapter 18

Orbital Eigenfunctions: 2-D

ORBITAL ANGULAR MOMENTUM EIGENFUNCTIONS

In the last lecture , we established that the operators $\vec{J}^2, J_z$ have a common set of eigenkets $|j,m\rangle$, $\vec{J}^2|j,m\rangle = j(j+1)\hbar^2|j,m\rangle$, $J_z|j,m\rangle = m\hbar|j,m\rangle$ where j, m are integers or half odd integers, and we found the matrix elements of J_+, J_- (and hence those of J_x, J_y) between these eigenkets. This purely formal structure, therefore, nails down the allowed values of total angular momentum and of any measured component. But there are other things we need to know: for example, how is an electron in a particular angular momentum state in an atom affected by an external field? To compute that, we need to know the wave function ψ.

If a system has spherical symmetry, such as an electron in the Coulomb field of a hydrogen nucleus, then the Hamiltonian H and the operators $\vec{J}^2$, J_z have a common set of eigenkets $|E,j,m\rangle$. The spherically symmetric Hamiltonian is unchanged by rotation, so must commute with any rotation operator, $\left[H,\vec{J}^2\right]=0$ and $\left[H,J_z\right]=0$. Recall that commuting Hermitian operators can be diagonalized simultaneously—and therefore have a common set of eigenkets.

Fortunately, many systems of interest do have spherical symmetry, at least to a good approximation, the basic example of course being the hydrogen atom, so the natural set of basis states is the common eigenkets of energy and angular momentum. It turns out that even when the spherical symmetry is broken, the angular momentum eigenkets may still be a useful starting point, with the symmetry breaking treated using perturbation theory.

TWO-DIMENSIONAL MODELS

As a warm-up exercise for the complications of the three-dimensional spherically symmetric model, it is worth analyzing a two-dimensional *circularly* symmetric model, that is,

$$H\psi(x,y)=-\frac{\hbar^2}{2M}\left(\frac{\partial^2}{\partial x^2}+\frac{\partial^2}{\partial y^2}\right)\psi(x,y)+V\left(\sqrt{(x^2+y^2)}\right)\psi(x,y)=E\psi(x,y).$$

(In this section, we'll denote the particle mass by *M*, to avoid confusion with the angular momentum quantum number *m* – but be warned you are often going to find *m* used for both in the same discussion!)

The two-dimensional angular momentum operator is

$$L=\vec{r}\times\vec{p}=xp_y-yp_x=-i\hbar\left(x\frac{\partial}{\partial y}-y\frac{\partial}{\partial x}\right).$$

It is a straightforward exercise to check that for the circularly-symmetric Hamiltonian above,

$[H,L]=0$

To take advantage of the circular symmetry, we switch to circular (r,ϕ), variables where

$$r=\sqrt{x^2+y^2},\quad \phi=\tan^{-1}(y/x),\ \text{so}\ x=r\cos\phi,\ y=r\sin\phi.$$

Transforming the Hamiltonian and angular momentum into (r,ϕ) coordinates,

$$H\psi(r,\phi) = -\frac{\hbar^2}{2M}\left(\frac{\partial^2}{\partial r^2}+\frac{1}{r}\frac{\partial}{\partial r}+\frac{1}{r^2}\frac{\partial^2}{\partial \phi^2}\right)\psi(r,\phi)+V(r)\psi(r,\phi)=E\psi(r,\phi)$$

and

$$L = -i\hbar\frac{\partial}{\partial\phi}.$$

The angular momentum eigenfunctions $\psi_m(r,\phi)$ satisfy

$$L\psi_m(r,\phi) = -i\hbar\frac{\partial}{\partial\phi}\psi_m(r,\phi) = m\hbar\,\psi_m(r,\phi)$$

equivalent to $L|m\rangle = m\hbar|m\rangle$. So $\psi_m(r,\phi) = R(r)e^{im\phi}$, and for this to be a single-valued wave function, *m* must be an integer. (This also ensures the hermiticity of the operator—the integration-by-parts check has canceling contributions from $\phi = 0$ and $\phi = 2\pi$.)

Notice this means that *any* function of *r* multiplied by $e^{im\phi}$ is an eigenfunction of angular momentum with eigenvalue $m\hbar$, and in fact any eigenfunction of *L* with eigenvalue must be of this form. So we can factor out the *r*-dependence, and write a complete set of orthonormal eigenfunctions of *L*, normalized by integrating around the circle:

$$\Phi_m(\phi) = \frac{1}{\sqrt{2\pi}}e^{im\phi}, \quad \phi \text{ an integer.}$$

It is interesting to note that this would be a complete set of wave functions for a particle confined to a ring—rather like the original Bohr orbits. In fact, nanotech rings in which electrons have wave functions like this can now be manufactured. Note also that in such rings one can also have *real* wave functions $\sqrt{1/\pi}\sin m\phi, \sqrt{1/\pi}\cos m\phi$, which are still energy eigenstates, but *not* angular momentum eigenstates, since they are standing waves, linear superpositions of waves going around the ring in opposite directions.

The common eigenstates of the Hamiltonian and the angular momentum evidently have the form

$$\left|E,m\right\rangle=\psi_{E,m}(r,\phi)=R_{E,m}(r)\Phi_m(\phi).$$

We should emphasize that although the angular part of the wave function does not depend on the radial potential, the radial component $R_{E,m}(r)$ *does* depend on the angular momentum m. This becomes obvious on putting this $\psi_{E,m}(r,\phi)$ into the (r,ϕ) version of Schrodinger's equation,

$$-\frac{\hbar^2}{2M}\left(\frac{\partial^2}{\partial r^2}+\frac{1}{r}\frac{\partial}{\partial r}+\frac{1}{r^2}\frac{\partial^2}{\partial\phi^2}\right)R_{E,m}(r)\Phi_m(\phi)+V(r)R_{E,m}(r)\Phi_m(\phi)$$
$$=ER_{E,m}(r)\Phi_m(\phi)$$

noting that $\partial^2/\partial\phi^2\equiv-m^2$, and canceling out the common factor $\Phi_m(\phi)$ to give

$$-\frac{\hbar^2}{2M}\left(\frac{d^2}{dr^2}+\frac{1}{r}\frac{d}{dr}-\frac{m^2}{r^2}\right)R_{E,m}(r)+V(r)R_{E,m}(r)=ER_{E,m}(r).$$

In this one-dimensional equation for the radial wave function $R_{E,m}(r)$, the angular momentum term $\hbar^2m^2/2Mr^2=L^2/2Mr^2$ evidently is equivalent to a repulsive potential. It's called the "centrifugal barrier" and is easy to understand from classical mechanics. To see this, consider a classical particle bound (in two dimensions) by an attractive central force $V(r)$. Split the momentum into a radial component p_r and a component in the direction perpendicular to the radius, $p_\perp$. The angular momentum $L=rp_\perp$ and is constant (since the force is central). The energy

$$E=\frac{p_r^2}{2M}+\frac{p_\perp^2}{2M}+V(r)=\frac{p_r^2}{2M}+\frac{L^2}{2Mr^2}+V(r)$$

substituting $p_\perp=L/r$. Since $L=m\hbar$, the angular part is exactly equivalent to the above Schrodinger equation.

But what about the radial part? Why isn't p_r just equal to $-i\hbar\partial/\partial r$, and p_r^2 equal to $\hbar^2\partial^2/\partial r^2$? We know the more complicated differentiation with respect to r in the Schrodinger equation above must be correct, because it came from $\partial^2/\partial x^2 + \partial^2/\partial y^2$ and $r = \sqrt{x^2+y^2}$.

To see why p_r equal to $-i\hbar\partial/\partial r$ is incorrect, even though it satisfies $[r, p_r] = i\hbar$, recall what happens in x-space. We argued there that $p_x = -i\hbar\partial/\partial x$ for a plane wave because from the photon analogy, acting on the plane wave state $Ce^{ip_x x/\hbar}$ this operator gives the rate of change of phase and therefore the momentum. But a *radial* wave is a little different: picture a photon wave coming out of a single slit having width far less than the photon wavelength. It will radiate outwards with equal amplitude in all directions (180°) but the *wave amplitude will decrease* with distance from the slit to conserve probability. For a long (narrow) slit, this is essentially a two-dimensional problem, so the wave function will be $\psi(r) \cong Ce^{ipr/\hbar}/\sqrt{r}$. We know that if we measure the momentum of photons at different distances from the slit we'll get the same result, the wavelength isn't changing, and that determines the phase behaviour. However, the operator $-i\hbar\partial/\partial r$ picks up *an extra term* from differentiating the $\sqrt{r}$, so it is obviously *not* giving us the momentum. Fortunately, this is easy to fix: we define the operator

$$\hat{p}_r = -i\hbar\left(\frac{\partial}{\partial r} + \frac{1}{2r}\right)$$

which eliminates the extra term, *and still satisfies* $[r, p_r] = i\hbar$.

However, there is still a small problem. If we substitute this $\hat{p}_r$ in the classical expression for the energy, following the procedure we used successfully to find Schrodinger's equation in Cartesian coordinates, we find

$$H = \frac{p_r^2}{2M} + \frac{L^2}{2Mr^2} + V(r)$$

$$= \frac{-\hbar^2\left(\frac{\partial}{\partial r} + \frac{1}{2r}\right)^2}{2M} + \frac{L^2}{2Mr^2} + V(r)$$

$$= \frac{-\hbar^2\left(\frac{\partial^2}{\partial r^2} + \frac{1}{r}\cdot\frac{\partial}{\partial r} - \frac{1}{4r^2}\right)}{2M} + \frac{L^2}{2Mr^2} + V(r)$$

This is almost—but not quite—the same as the equation we found by transforming from Cartesian coordinates. The difference is the term $\hbar^2/8Mr^2$. So which is right? Actually our first one was right—this second one, derived directly from the classical Hamiltonian, does give the same result in the classical limit, because the difference between them vanishes for $\hbar \to 0$. We conclude that beginning with the classical Hamiltonian, and replacing dynamical variables with the appropriate quantum operators, cannot guarantee that we get the correct quantum Hamiltonian: it might be off by some term of order $\hbar$. This would become evident in predicting properties of truly quantum systems, such as atomic energy levels. Problems of this kind are common in constructing quantum theories starting from a classical theory: essentially, in a classical theory, the order of variables in an expression is irrelevant, but in the quantum theory there can only be *one* correct order of noncommuting variables such as $\partial/\partial r$ and r in any expression.

What can we say about the radial wave function $R_{E,m}(r)$? If both the energy and the potential at the origin are finite, then for small r $R_{E,m}(r) \approx Ar^m$ or Ar^{-m}. However, the wave function cannot be discontinuous, so $R_{E,m}(r) \approx Ar^{|m|}$. To make further progress in finding the wave function, we need to know the potential. Specific examples will be analysed in due course. It is interesting to note that the allowed wave functions, proportional to $r^m e^{im\phi}$, $r^m e^{-im\phi}$, $m > 0$, are the complex functions z^m, $\left(z^*\right)^m$ if the two-dimensional space is mapped into the complex plane. Representing many-electron wave functions in the plane in this way was a key to understanding the quantum Hall effect.

Chapter 19

Orbital Eigenfunctions in 3-D

THE ANGULAR MOMENTUM OPERATORS IN SPHERICAL POLAR COORDINATES

The angular momentum operator $\vec{L} = \vec{r} \times \vec{p} = -i\hbar\vec{r} \times \vec{\nabla}$.

In spherical polar coordinates,

$$x = r \sin\theta \cos\phi$$

$$y = r \sin\theta \sin\phi$$

$$z = r \cos\theta$$

$$ds^2 = dr^2 + r^2 d\theta^2 + r^2 \sin^2\theta d\phi^2$$

the gradient operator is

$$\vec{\nabla} = \hat{r}\frac{\partial}{\partial r} + \hat{\theta}\frac{1}{r}\frac{\partial}{\partial \theta} + \hat{\phi}\frac{1}{r\sin\theta}\frac{\partial}{\partial \phi}$$

where now *the little hats denote unit vectors*: $\hat{r}$ is radially outwards, $\hat{\theta}$ points along a line of longitude away from the north pole (and therefore in the direction of increasing θ) and $\hat{\phi}$ points along a line of latitude in an anticlockwise direction as seen looking down on the north pole (that is, in the direction of increasing ϕ). $\hat{r}$, $\hat{\theta}$, $\hat{\phi}$ form an orthonormal local basis, and

$$\hat{r} \times \hat{\theta} = \hat{\phi}, \quad \hat{r} \times \hat{\phi} = -\hat{\theta}$$

So

$$\vec{r}\times\vec{\nabla} = \hat{\phi}\frac{\partial}{\partial\theta} - \hat{\theta}\frac{1}{\sin\theta}\frac{\partial}{\partial\phi}.$$

(Explicitly, $\hat{\phi} = (-\sin\phi, \cos\phi, 0)$ and $\hat{\theta} = (\cos\theta\cos\phi, \cos\theta\sin\phi, -\sin\theta)$

The vector $\hat{\phi}$ has zero component in the z-direction, the vector $\hat{\theta}$ has component $-\sin\theta$ in the z-direction, so we can immediately conclude that

$$L_z = (\vec{r}\times\vec{p})_z = -(i\hbar\vec{r}\times\vec{\nabla})_z = -i\hbar\frac{\partial}{\partial\phi}$$

just as in the two-dimensional case.

The operator

$$L^2 = -\hbar^2\left(\hat{\phi}\frac{\partial}{\partial\theta} - \hat{\theta}\frac{1}{\sin\theta}\frac{\partial}{\partial\phi}\right)\cdot\left(\hat{\phi}\frac{\partial}{\partial\theta} - \hat{\theta}\frac{1}{\sin\theta}\frac{\partial}{\partial\phi}\right).$$

To evaluate this expression, we use $\hat{\phi}^2 = 1$, $\hat{\theta}^2 = 1$, $\hat{\phi}\cdot\hat{\theta} = 0$ but we must also check the effects of the differential operators in the first expression on the variables in the second, *including the unit vectors*. Notice $\partial\hat{\phi}/\partial\theta = 0$, $\partial\hat{\theta}/\partial\theta$ is in the r-direction, $\partial\hat{\phi}/\partial\phi$ is a horizontal unit vector pointing inwards perpendicular to $\hat{\phi}$, and having component $-\cos\theta$ in the $\hat{\theta}$-direction, $\partial\hat{\theta}/\partial\phi = \hat{\phi}\cos\theta$. Therefore, the *only* "differentiation of a unit vector" term that contributes to L^2 is $\hbar^2\hat{\theta}\frac{1}{\sin\theta}\cdot\frac{\partial\hat{\phi}}{\partial\phi}\frac{\partial}{\partial\theta} = -\hbar^2\cot\theta\cdot\frac{\partial}{\partial\theta}$. The $\hat{\phi}\frac{\partial}{\partial\theta}$ acting on the $\sin\theta$ in $-\hat{\theta}\frac{1}{\sin\theta}\frac{\partial}{\partial\phi}$ contributes nothing because $\hat{\phi}\cdot\hat{\theta} = 0$.

Therefore

$$L^2 = -\hbar^2\left(\frac{\partial^2}{\partial\theta^2} + \cot\theta\frac{\partial}{\partial\theta} + \frac{1}{\sin^2\theta}\frac{\partial^2}{\partial\phi^2}\right)$$

$$= -\hbar^2\left(\frac{1}{\sin\theta}\frac{\partial}{\partial\theta}\sin\theta\frac{\partial}{\partial\theta}+\frac{1}{\sin^2\theta}\frac{\partial^2}{\partial\phi^2}\right)$$

Now, we know that L^2 and L_z have a common set of eigenkets (since they commute) and we've already established that those of L_z are $\Phi(\phi)=e^{im\phi}/\sqrt{2\pi}$, with m an integer, so the eigenkets of L^2 must have this same *f* dependence, so they must be of the form $\Theta_l^m(\theta)\Phi(\phi)$, where $\Theta_l^m(\theta)$ is a (suitably normalized) solution of the equation

$$\frac{1}{\sin\theta}\frac{d}{d\theta}\sin\theta\frac{d\Theta_l^m(\theta)}{d\theta}-\frac{m^2}{\sin^2\theta}\Theta_l^m(\theta)=-l(l+1)\Theta_l^m(\theta)$$

more conveniently written

$$\sin\theta\frac{d}{d\theta}\sin\theta\frac{d\Theta_l^m(\theta)}{d\theta}+\left(l(l+1)\sin^2\theta-m^2\right)\Theta_l^m(\theta)=0.$$

Finding the $m = l$ Eigenket of L^2, L_z

Recall now that for the simple harmonic oscillator, the easiest wave function to find was that of the ground state, the solution of the linear equation $\hat{a}\psi_0=0$, and the other state wave functions could then be found by applying the creation operator in differential form the necessary number of times.

A similar strategy works here: we find the *highest* state on the l ladder, $m = l$, the state $|l,l\rangle$, from the equation $L_+|l,l\rangle=0$, where $L_+=L_x+iL_y$. So we need to find $L_+=-i\hbar\left(\vec{r}\times\vec{\nabla}\right)_+$.

From $\vec{r}\times\vec{\nabla}=\hat{\phi}\frac{\partial}{\partial\theta}-\hat{\theta}\frac{1}{\sin\theta}\frac{\partial}{\partial\phi}$,

we have

$$\left(\vec{r}\times\vec{\nabla}\right)_+=\hat{\phi}_+\frac{\partial}{\partial\theta}-\hat{\theta}_+\frac{1}{\sin\theta}\frac{\partial}{\partial\phi},$$

and using $\hat{\phi} = (-\sin\phi, \cos\phi, 0)$, $\hat{\theta} = (\cos\theta\cos\phi, \cos\theta\sin\phi, -\sin\theta)$ we see that $\hat{\phi}_+$, the component of $\hat{\phi}$ in the + direction, is $\hat{\phi}_+ = \hat{\phi}_x + i\hat{\phi}_y = ie^{i\phi}$, and similarly $\hat{\theta}_+ = \cos\theta e^{i\phi}$.

Therefore,

$$L_+ = \hbar e^{i\phi}\left(\frac{\partial}{\partial\theta} + i\cot\theta\frac{\partial}{\partial\phi}\right)$$

$$L_- = -\hbar e^{-i\phi}\left(\frac{\partial}{\partial\theta} - i\cot\theta\frac{\partial}{\partial\phi}\right).$$

So $L_+|l,l\rangle = 0$ becomes

$$\left(\frac{\partial}{\partial\theta} + i\cot\theta\frac{\partial}{\partial\phi}\right)\Theta_l^l(\theta)e^{il\phi} = 0$$

that is,

$$\left(\frac{d}{d\theta} - l\cot\theta\right)\Theta_l^l(\theta) = 0.$$

The solution to this equation is

$$\Theta_l^l(\theta) = N(\sin\theta)^l$$

where N is the normalization constant. The $m \neq l$ wave functions are generated by applying the lowering operator L.

NORMALIZING THE $m = l$ EIGENKET

The standard notation for the normalized eigenkets $|l,m\rangle$ is $Y_l^m(\theta,\phi) = \Theta_l^m(\theta)\Phi_m(\phi)$. These functions, being eigenkets of Hermitian operators with different eigenvalues, must satisfy

$$\int_{\theta=0}^{\pi}\int_{\phi=0}^{2\pi} Y_{l'}^{m'*}(\theta,\phi)Y_l^m(\theta,\phi)\sin\theta d\theta d\phi = \int Y_{l'}^{m'*}(\theta,\phi)Y_l^m(\theta,\phi)d\Omega = \delta_{ll'}\delta_{mm'}.$$

So, our first job is to normalize $\Theta_l^l(\theta) = N(\sin\theta)^l$ (taking $\Phi_l(\phi) = e^{il\phi}/\sqrt{2\pi}$ already normalized).

$$|N|^2 \int_0^{\pi} (\sin\theta)^{2l+1} d\theta = 1$$

The integral can be evaluated using the substitution $\mu = \cos\theta$ to give $\int_{-1}^{1}(1-\mu^2)^l d\mu$, then making the further substitution $u = \frac{1}{2}(1-\mu)$ to give $2^{2l+1}\int_0^1 u^l(1-u)^l du$, which can be integrated by parts to give

$$|N|^2\, 2^{2l+1}(l!)^2/(2l+1)! = 1.$$

Therefore

$$Y_l^l(\theta,\phi) = (-1)^l\left(\frac{(2l+1)!}{4\pi}\right)^{1/2}\frac{1}{2^l l!}(\sin\theta)^l e^{il\phi} = c_l(\sin\theta)^l e^{il\phi}$$

where we have fixed the sign in accord with the standard convention, and we will denote the rather cumbersome normalization constant by c_l.

Notice that for large values of l, this function is heavily weighted around the equator, as we would expect—for a given total angular momentum one gets a maximum component in the *z*-direction when the motion is concentrated in the *x*, *y* plane. This looks like a Bohr orbit.

FINDING THE REST OF THE EIGENKETS

Now that $|l,l\rangle$ is normalized, we can automatically produce correctly normalized $|l,m\rangle$'s, since we know the matrix element of the lowering operator between normalized states. We don't have to do any more integrals.

For example, $L_-|l,l\rangle = \hbar\sqrt{2l}\,|l,l-1\rangle$, equivalently (the $\hbar$'s of course cancel)

$$Y_l^{l-1}(\theta,\phi) = \frac{(-1)}{\sqrt{2l}}e^{-i\phi}\left(\frac{\partial}{\partial\theta} - i\cot\theta\frac{\partial}{\partial\phi}\right)Y_l^l.$$

That is,

$$Y_l^{l-1}(\theta,\phi) = c_l e^{-i\phi}\left(\frac{\partial}{\partial\theta} - i\cot\theta\frac{\partial}{\partial\phi}\right)\sin^l\theta\cdot e^{il\phi}$$

$$= -c_l e^{i(l-1)\phi}\sqrt{2l}\sin^{l-1}\theta\cos\theta$$

(both terms giving equal contributions).

Note that this function is actually *zero* on the equator, but for large l it peaks close to the equator (on both sides).

In principle, we can reapply this differential operator over and over to generate all the $|l,m\rangle$ states, but this gets very messy. However, there is a neat theorem concerning the lowering operator that makes it all straightforward:

$$L_- e^{im\phi}f(\theta) = e^{i(m-1)\phi}\left(\sin^{1-m}\theta\frac{d}{d(\cos\theta)}\sin^m\theta\right)f(\theta)$$

So

$$L_- e^{il\phi}\sin^l\theta = e^{i(l-1)\phi}\left(\sin^{1-l}\theta\frac{d}{d(\cos\theta)}\sin^l\theta\right)\sin^l(\theta)$$

and applying the operator again,

$$(L_-)^2 e^{il\phi}\sin^l\theta = L_- e^{i(l-1)\phi}\left(\sin^{1-l}\theta\frac{d}{d(\cos\theta)}\sin^l\theta\right)\sin^l(\theta)$$

$$= e^{i(l-2)\phi}\left(\sin^{2-l}\theta\frac{d}{d(\cos\theta)}\sin^{l-1}\theta\right)\cdot\left(\sin^{1-l}\theta\frac{d}{d(\cos\theta)}\sin^l\theta\right)\sin^l(\theta)$$

$$= e^{i(l-2)\phi}\left(\sin^{2-l}\theta\frac{d^2}{d^2(\cos\theta)}\sin^l\theta\right)\sin^l(\theta).$$

So the point of introducing this odd-looking representation of the lowering operator is that the $\sin^{1-l}\theta$ term in the middle is exactly *canceled* when the operator is applies twice, and similar cancellations occur on repeating the operation, giving the (relatively) simple representation:

$$Y_l^m(\theta,\phi) = c_l\sqrt{\frac{(l+m)!}{(2l)!(l-m)!}}e^{im\phi}\sin^{-m}\theta\frac{d^{l-m}}{d(\cos\theta)^{l-m}}\sin^{2l}\theta$$

(Where did all those factorials come from? They're the product of all the inverse square root factors in $|l,m-1\rangle = \frac{1}{\sqrt{(l+m)(l-m+1)}} L_{-}|l,m\rangle$ for the number of lowerings necessary.)

Note that for $m = 0$ the function is

$$Y_l^0(\theta,\phi) = c_l \sqrt{\frac{1}{(2l)!}} \frac{d^l}{d(\cos\theta)^l} \sin^{2l}\theta,$$

and in fact not a function of ϕ at all. This isn't surprising, since it has zero angular momentum about the z-direction, the appropriate $\Phi(\phi)$ is just constant.

For $m = -l$, the differentiation becomes trivial, because, writing $\cos\theta = \mu$, the differentiation becomes $\frac{d^{2l}}{d\mu^{2l}}\left(1-\mu^2\right)^l$ and only the μ^{2l} term survives, giving

$$Y_l^{-l}(\theta,\phi) = (-1)^l c_l e^{-il\phi} \sin^l\theta.$$

Of course, this could also have been found from the linear equation $L_{-}|l,-l\rangle = 0$, and we could have equally generated all the states by applying L_{+} to this state. In fact, this gives a different—but of course equivalent—expression for the $Y_l^m(\theta,\phi)$:

$$Y_l^m(\theta,\phi) = (-1)^m c_l \sqrt{\frac{(l-m)!}{(2l)!(l+m)!}} e^{im\phi} \sin^m\theta \frac{d^{l+m}}{d(\cos\theta)^{l+m}} \sin^{2l}\theta$$

RELATING THE Y_l^m'S TO THE LEGENDRE FUNCTIONS

The *Legendre polynomials* $P_n(\cos\theta)$ are defined by:

$$P_n(\cos\theta) = \frac{1}{2^n n!} \frac{d^n}{d(\cos\theta)^n} \sin^{2n}\theta, \text{ or}$$

$$P_n(\mu) = \frac{1}{2^n n!}\frac{d^n}{d\mu^n}\left(1-\mu^2\right)^n$$

where $\mu = \cos\theta$, so $d\mu = -\sin\theta d\theta$. From this form, it is easy to show that $P_n(1) = 1$ (all n differentiations must take out a $\left(1-\mu^2\right)$ factor to give a nonzero contribution), and $P_n(\mu)$ must have n zeros in the interval (-1, 1). alternates between an even function and an odd function.

The normalization of the 's is

$$\int_{-1}^{1}\left(P_n(\mu)\right)^2 d\mu = \left(\frac{1}{2^n n!}\right)^2 \int_{-1}^{1}\frac{d^n}{d\mu^n}\left(\mu^2-1\right)^n \frac{d^n}{d\mu^n}\left(\mu^2-1\right)^n d\mu$$

$$= (-1)^n\left(\frac{1}{2^n n!}\right)^2 \int_{-1}^{1}\left(\mu^2-1\right)^n \frac{d^{2n}}{d\mu^{2n}}\left(\mu^2-1\right)^n d\mu$$

$$= (2n)!\left(\frac{1}{2^n n!}\right)^2 \int_{-1}^{1}\left(\mu^2-1\right)^n d\mu$$

$$= \frac{2}{2n+1}$$

where in that last line we used the result for the integral obtained earlier in this lecture for normalizing Y_l^l.

Doing the same repeated integration by parts for two different Legendre polynomials proves they are orthogonal,

$$\int_{-1}^{1} P_m(\mu)P_n(\mu)d\mu = 0, \quad m \neq n.$$

The *associated Legendre functions* are defined (for *n* and *m* zero or positive integers, $n \geq m$) by:

$$P_n^m(\mu) = \left(1-\mu^2\right)^{m/2}\frac{d^m}{d\mu^m}P_n(\mu)$$

$$= (-1)^n\frac{\left(1-\mu^2\right)^{m/2}}{2^n n!}\frac{d^{n+m}}{d\mu^{n+m}}\left(1-\mu^2\right)^n.$$

Following Messiah in requiring $Y_l^0(0,0)$ be real and positive, we find

$$Y_l^0(\theta,\phi)=\sqrt{\frac{2l+1}{4\pi}}P_l(\cos\theta)$$

where the coefficient just reflects the differing normalization conventions. Similarly, the spherical harmonics with nonzero *m* are proportional to the associated Legendre functions (the odd ones are *not* polynomials in $\cos\theta$, despite Shankar p. 337, since they include odd powers of $\sin\theta$),

$$Y_l^m(\theta,\phi)=\sqrt{\frac{(2l+1)(l-m)!}{4\pi(l+m)!}}(-1)^m e^{im\phi}P_l^m(\cos\theta).$$

THE SPHERICAL HARMONICS AS A BASIS

We have found explicit expressions for the spherical harmonics: an orthonormal set of eigenfunctions of L^2 and L_z defined on the surface of a sphere,

$$\int_{\theta=0}^{\pi}\int_{\phi=0}^{2\pi}Y_{l'}^{m'*}(\theta,\phi)Y_l^m(\theta,\phi)\sin\theta d\theta d\phi=\int Y_{l'}^{m'*}(\theta,\phi)Y_l^m(\theta,\phi)d\Omega=\delta_{l'l}\delta_{m'm}.$$

They form a *complete* set:

$$\sum_{l=0}^{\infty}\sum_{m=-l}^{l}|l,m\rangle\langle l,m|=I$$

or

$$\sum_{l=0}^{\infty}\sum_{m=-l}^{l}Y_l^{m*}(\theta,\phi)Y_l^m(\theta',\phi')=\delta(\cos\theta-\cos\theta')\delta(\phi-\phi')=\delta(\Omega-\Omega')$$

in the notation of Messiah, where W refers to a point on the spherical surface.

(Formal proof of the completeness is given in Byron and Fuller, *Mathematics of Classical and Quantum Physics*.)

The above equation could also be written

$$\sum_{l=0}^{\infty}\sum_{m=-l}^{l}\langle\theta,\phi|l,m\rangle\langle l,m|\theta',\phi'\rangle=\langle\theta,\phi|\theta',\phi'\rangle$$

$$=\delta(\cos\theta-\cos\theta')\delta(\phi-\phi')$$

where the ket $|\theta',\phi'\rangle$ is to be understood as a localized ket, the spherical-surface version of $|x\rangle$, normalized by its *d*-function inner product with the bra $\langle\theta,\phi|$, exactly analogous to $\langle x|x'\rangle = \delta(x-x')$, bearing in mind that the infinitesimal area element is $-d(\cos\theta)d\phi$, (a positive quantity in the relevant interval, 0 to π).

This completeness means that any reasonable function on the surface of the sphere can be expressed as a sum over spherical harmonics with appropriate coefficients, in other words the spherical generalization of a Fourier series.

In fact, L^2 is equivalent to ∇^2 on the spherical surface, so the Y_l^m are the eigenfunctions of the operator ∇^2. Just as in one dimension the eigenfunctions of d^2/dx^2 have the spatial dependence of the eigenmodes of a vibrating string, the spherical harmonics have the spatial dependence of the eigenmodes of a vibrating spherical balloon. Of course, to describe the displacement of the balloon skin (which must be real!) with these eigenfunctions, we can no longer use the eigenfunctions of the *z*-component of angular momentum, since they are complex except in the trivial zero case. We must rearrange the eigenfunctions of L^2, for example replacing the pair $e^{i\phi}, e^{-i\phi}$ with $\cos\phi$, $\sin\phi$. These real solutions, essentially $\frac{1}{\sqrt{2}}\left(|l,l\rangle \pm |l,-l\rangle\right)$, have *l* nodal lines (zeroes) of longitude. Moving down one notch in $|m|$, the (real) state with $|m| = l-1$ has $l-1$ longitudinal nodes, but has added a *latitudinal* node: the equator. Then $|m| = l-2$ has $l-2$ longitudinal nodes, 2 latitudinal nodal lines—*there are always l nodal lines total.*

Some of these modes of vibration have been observed in the sun after a sunspot storm. The spherical harmonics are also used in analyzing the cosmic background radiation.

SOME LOW ORDER SPHERICAL HARMONICS

Let's look in more detail at the lowest order spherical harmonics, from our general formulas in the previous lecture.

$$Y_0^0 = \frac{1}{\sqrt{4\pi}}$$

$$Y_1^1 = -\sqrt{\frac{3}{8\pi}} \sin\theta e^{i\phi}$$

$$Y_1^0 = \sqrt{\frac{3}{4\pi}} \cos\theta$$

$$Y_1^{-1} = \sqrt{\frac{3}{8\pi}} \sin\theta e^{-i\phi}$$

$$Y_2^2 = \sqrt{\frac{15}{32\pi}} \sin^2\theta e^{2i\phi}, \quad Y_2^1 = -\sqrt{\frac{15}{8\pi}} \sin\theta\cos\theta e^{i\phi}, \quad Y_2^0 = \sqrt{\frac{5}{16\pi}}\left(3\cos^2\theta - 1\right)$$

$$Y_2^{-2} = \sqrt{\frac{15}{32\pi}} \sin^2\theta e^{-2i\phi}, \quad Y_2^{-1} = \sqrt{\frac{15}{8\pi}} \sin\theta\cos\theta e^{-i\phi}$$

It is often useful to write the Y_l^m in terms of Cartesian coordinates,

$$(x,y,z) = (r\sin\theta\cos\phi,\ r\sin\theta\sin\phi,\ r\cos\theta)$$

so

$$Y_1^1(x,y,z) = -\sqrt{\frac{3}{8\pi}} \cdot \frac{x+iy}{r}, \quad Y_1^0(x,y,z) = \sqrt{\frac{3}{4\pi}} \cdot \frac{z}{r}, \quad Y_1^{-1}(x,y,z) = \sqrt{\frac{3}{8\pi}} \cdot \frac{x-iy}{r}$$

and

$$Y_2^2 = \sqrt{\frac{15}{32\pi}}(x+iy)^2, \quad Y_2^1 = -\sqrt{\frac{15}{8\pi}}(x+iy)z, \quad Y_2^0 = \sqrt{\frac{5}{16\pi}}(3z^2-1), \text{ etc.}$$

THE Y_1^m AS A BASIS OF THE $l = 1$ SUBSPACE

The Y_1^m are the $l = 1$ eigenstates of L^2 and L_z. But what if we'd chosen to look for the common eigenstates of L^2 and L_x instead? What $l = 1$ state has zero angular momentum component in the direction of the x-axis? Clearly it will be $\sqrt{\frac{3}{4\pi}}\cdot\frac{x}{r}$, in other words the previous Y_1^0 with z replaced by x, because after all, our labeling of axes was arbitrary.

Now,

$$\sqrt{\frac{3}{4\pi}}\cdot\frac{x}{r} \text{ is just } \left(1/\sqrt{2}\right)\left(-Y_1^1+Y_1^{-1}\right).$$

In fact, *any* $l = 1$ state, with a specified component in *any* direction, can be written as

$$\alpha_1\left|1,1\right\rangle+\alpha_0\left|1,0\right\rangle+\alpha_{-1}\left|1,-1\right\rangle=\sum\alpha_m\left|1,m\right\rangle.$$

This can be seen as follows: an $l = 1$ state has to be *linear* in x/r, y/r, z/r (any quadratic term would give rise to $e^{2i\phi}$ about an appropriate axis, call that the z-axis, so $m = 2$ and l must be 2 or greater), and any such state can be written as a linear combination of

$$(x+iy)/r,\ (x-iy)/r,\ z/r.$$

The bottom line, then, is that the Y_1^m do indeed provide a *complete* basis for the $l = 1$ space of eigenstates of L^2.

REPRESENTING THE ROTATION OPERATOR WITHIN THE $L = 1$ SUBSPACE

Recall that we originally introduced the angular momentum operator $\vec{J}$ by defining it as the generator of infinitesimal rotations when acting on any wave function, including multicomponent wave functions. We found, using the commutativity properties of ordinary rotations, that the vector components of $\vec{J}$ had to satisfy $\left[J_x,J_y\right]=i\hbar J_z$, etc., and from that we deduced the possible sets of eigenvalues of the commuting pair of operators $\vec{J}^2$, J_z were $j(j+1)\hbar^2$ for $\vec{J}^2$, with j an integer of half an odd integer, and for each such j the allowed eigenvalues of J_z were $m\hbar, \quad m=-j,-j+1,\ldots,+j$.

Back to the l = 1 angular wave functions: we have established that any such function can be written $\alpha_1|1,1\rangle+\alpha_0|1,0\rangle+\alpha_{-1}|1,-1\rangle=\sum\alpha_m|1,m\rangle$, and so is a vector in a three-dimensional space spanned by the set $|1,m\rangle$, $m=1,0,-1$. In other words, the wave function is a three-component object. The angular momentum operator must therefore be *a matrix operator in this three-dimensional space,* such that, by definition, the effect of an infinitesimal rotation on the multicomponent wave function is:

$$R\left(\delta\vec{\theta}\right)\psi_{l=1}(\theta,\phi)=e^{-\frac{i}{\hbar}\delta\vec{\theta}\cdot\hat{J}}\begin{pmatrix}\alpha_1\\ \alpha_0\\ \alpha_{-1}\end{pmatrix}=\begin{pmatrix}\alpha_1'\\ \alpha_0'\\ \alpha_{-1}'\end{pmatrix}$$

The unitary rotation operator acting in the l = 1 subspace, , has to be a matrix. The standard notation for its matrix elements is:

The unitary rotation operator acting in the l = 1 subspace, $U\left(R\left(\vec{\theta}\right)\right)=e^{-\frac{i\vec{\theta}\cdot\vec{J}}{\hbar}}$, has to be a 3×3 matrix. The standard notation for its matrix elements is:

$$D^{(1)}_{m'm}\left(R\left(\vec{\theta}\right)\right)=\langle 1,m'|e^{-\frac{i\vec{\theta}\cdot\vec{J}}{\hbar}}|1,m\rangle$$

so the rotated ket is

$$\alpha'_{m'}=\sum_{m,m}D^{(1)}_{m'm}\alpha_m,\quad \text{or } \alpha'=D\alpha$$

To evaluate this matrix explicitly, we must expand the exponential and we need the matrix elements of J_z, J_+, J_- between the states $|1,m\rangle$ which we already know.

Now, the basis of the three-dimensional space is just the common eigenkets of $\vec{J}^2$, J_z, in this case identical to $\vec{L}^2$, L_z. We know the matrix elements of between states from the earlier lecture, so it is simple to find the matrix representations of the components of J in this space:

$$J_x^{(1)} = \frac{\hbar}{\sqrt{2}}\begin{pmatrix} 0 & 1 & 0 \\ 1 & 0 & 1 \\ 0 & 1 & 0 \end{pmatrix}, \quad J_y^{(1)} = \frac{i\hbar}{\sqrt{2}}\begin{pmatrix} 0 & -1 & 0 \\ 1 & 0 & -1 \\ 0 & 1 & 0 \end{pmatrix},$$

$$J_z^{(1)} = \hbar\begin{pmatrix} 1 & 0 & 0 \\ 0 & 0 & 0 \\ 0 & 0 & -1 \end{pmatrix}.$$

We have added the superscript (1) because this representation of the infinitesimal rotation operators is specific to j = 1 (representations for general values of j are as $(2j+1)\times(2j+1)$ matrices, reflecting the dimensionality of the space spanned by the $2j$ + 1 distinct m values).

Expanding the exponential is not difficult, because by inspection $\left(J_z^{(1)}/\hbar\right)^3 = \left(J_z^{(1)}/\hbar\right)$, so from spherical symmetry $\left(\hat{\vec{n}}\cdot\vec{J}^{(1)}/\hbar\right)^3 = \left(\hat{\vec{n}}\cdot\vec{J}^{(1)}/\hbar\right)$ for a unit vector in any direction. The result is:

$$D^{(1)}\left(R(\vec{\theta})\right) = e^{-\frac{i\theta\hat{n}\cdot\vec{J}}{\hbar}} = I + (\cos\theta - 1)\left(\frac{\hat{\vec{n}}\cdot\vec{J}}{\hbar}\right)^2 - i\sin\theta\left(\frac{\hat{\vec{n}}\cdot\vec{J}}{\hbar}\right).$$

Chapter 20

The Density Matrix

PURE STATES AND MIXED STATES

Our treatment here more or less follows that of Sakurai, beginning with two imagined Stern-Gerlach experiments. In that experiment, a stream of (non-ionized) silver atoms from an oven is directed through an inhomogeneous vertical magnetic field, and the stream splits into two. The silver atoms have nonzero magnetic moments, and a magnetic moment in an inhomogeneous magnetic field experiences a nonzero force, causing the atom to veer from its straight line path, the magnitude of the deflection being proportional to the component of the atom's magnetic moment in the vertical (field) direction. The observation of the beam splitting into two, and no more, means that the vertical component of the magnetic moment, and therefore the associated angular momentum, can only have *two* different values. From the basic analysis of rotation operators and the properties of angular momentum that follow, this observation forces us to the conclusion that the total angular momentum of a silver atom is $\frac{1}{2}\hbar$. Ordinary orbital angular momenta cannot have half-integer values; this experiment was one of the first indications that the electron has a spin degree of freedom, an angular momentum that cannot be interpreted as orbital angular momentum of constituent parts. The silver atom has 47 electrons, 46 of them have total spin and orbital momenta that separately cancel, the 47th has no orbital angular momentum, and its spin is the entire angular momentum of the atom.

Here we shall use the Stern-Gerlach stream as an example of a large collection of quantum systems (the atoms) to clarify just how to describe such a collection, often called an *ensemble*. To avoid unnecessary complications, we only consider the *spin* degrees of freedom. We begin by examining two different streams:

Suppose experimentalist *A* prepares a stream of silver atoms such that each atom is in the spin state ψ_A:

$$|\psi_A\rangle = \frac{1}{\sqrt{2}}\left(|\uparrow\rangle + |\downarrow\rangle\right)$$

Meanwhile, experimentalist *B* prepares a stream of silver atoms which is a *mixture*: half the atoms are in state $|\uparrow\rangle$ and half are in the state $|\downarrow\rangle$: call this mix *B*.

Question: can we distinguish the *A* stream from the *B* stream?

Evidently, not by measuring the spin in the *z*-direction! Both will give up 50per cent of the time, down 50per cent. But: we *can* distinguish them by measuring the spin in the *x*-direction: the ψ_A quantum state is in fact just that of a spin in the *x*-direction, so it will give "up" in the *x*-direction every time from now on we call it $|\uparrow_x\rangle$, whereas the state $|\uparrow\rangle$ ("up" in the *z*-direction) will yield "up" in the *x*-direction only 50per cent of the time, as will $|\downarrow\rangle$.

The state $\psi_A = |\uparrow_x\rangle$ is called a *pure* state, it's the kind of quantum state we've been studying this whole course.

The stream *B*, in contrast, is in a *mixed* state: the kind that actually occurs to a greater or lesser extent in a real life stream of atoms, different pure quantum states occurring with different probabilities, but with no phase coherence between them. In other words, these relative probabilities in *B* of different quantum states do *not* derive from probability amplitudes, as they do in finding the probability of spin up in stream *A*: the probabilities of the different quantum states in the mixed state *B* are exactly like classical probabilities.

That being said, though, to find the probability of measuring spin up in some such mixed state, one *first* uses the classical-type probability for each component state, *then* for each quantum state in the mix, one finds the probability of spin up *in that state* by the standard quantum technique.

Theerefore, for a mixed state in which the system is in state $|\psi_i\rangle$ with probability w_i, $\Sigma w_i = 1$ the expectation value of an operator A is

$$\langle A\rangle = \sum w_i \langle\psi_i|A|\psi_i\rangle$$

and we should emphasize that these $|\psi_i\rangle$ do *not* need to be orthogonal (but they are of course normalized): for example one could be $|\uparrow_x\rangle$, another $|\uparrow_z\rangle$. (We put the usually omitted z in for emphasis.)

THE DENSITY MATRIX

The equation for the expectation value <A> can be written:

$$\langle A\rangle = \text{Trace}(\hat{\rho}A) \text{ where } \hat{\rho} = \sum w_i |\psi_i\rangle\langle\psi_i|.$$

To see exactly how this comes about, recall that for an operator B in a finite-dimensional vector space with an orthonormal basis set , $TrB = \sum_{j=1}^{n} \langle j|B|j\rangle = B_{jj}$, where the repeated suffix implies summation of the diagonal matrix elements of the operator. Therefore,

$$\begin{aligned}\text{Tr}(\hat{\rho}A) &= \sum_{j=1}^{n}\sum_{i=1}^{n} \langle j|w_i|\psi_i\rangle\langle\psi_i|A|j\rangle \\ &= \sum_{j=1}^{n}\sum_{i=1}^{n} \langle\psi_i|A|j\rangle\langle j|w_i|\psi_i\rangle \\ &= \sum_{i=1}^{n} w_i \langle\psi_i|A|\psi_i\rangle\end{aligned}$$

since $\Sigma|j><j|=1$, the identity.This $\hat{\rho}$ is called the *density matrix*: its matrix form is made explicit by considering states $|\psi_i>$ in a finite N-dimensional vector space (such as spins or angular momenta) $|\psi_i> = \sum_j (v_i)_j |j>$ where the $|j>$ are an orthonormal basis set, and $(v_i)_j$ is the jth component of a normalized vector V_i. It is convenient to express $\hat{\rho}$ in terms of kets and bras belonging to this orthonormal basis,

$$\hat{\rho} = \sum w_i |\psi_i\rangle\langle\psi_i| = \sum_{i,j,k} w_i (V_i)_j (V_i^\dagger)_k |j\rangle\langle k| = \sum_{j,k} \rho_{jk} |j\rangle\langle k|$$

and evidently

$$\langle A\rangle = \text{Trace}(\hat{\rho}A) = \sum_{n,j,k} \langle n|\rho_{jk}|j\rangle\langle k|A|n\rangle = \sum_{j,k} \rho_{jk}\langle k|A|j\rangle = \sum_{j,k} \rho_{jk}A_{kj}.$$

Since ρ_{jk} is just a *number*, $\langle n|\rho_{jk}|j\rangle = \rho_{jk}\langle n|j\rangle = \rho_{jk}\delta_{nj}$.

$\text{Trace}(\hat{\rho}A)$ is *basis-independent*, the trace of a matrix being unchanged by a unitary transformation, since it follows from $\text{Tr}ABC = \text{Tr}BCA$ that $\text{Tr}\,U^\dagger AU = \text{Tr}\,AUU^\dagger = \text{Tr}\,A$ for $UU^\dagger = 1$.

Note that since the vectors V_i are normalized, $\sum_j (V_i)_j (V_i^\dagger)_j = 1$, with the i not summed over, and $\sum w_i = 1$, it follows that

$$Tr\hat{\rho} = 1$$

(also evident by putting $A = 1$ in the equation for <A>).

For a system in a *pure* quantum state $|\psi\rangle$, $\hat{\rho} = |\psi\rangle\langle\psi|$, just the projection operator into that state, and

$$\hat{\rho}^2 = \hat{\rho},$$

as for all projection operators.

It's worth spelling out how this differs from the mixed state by looking at the form of the density matrix.

For the pure state $|\psi\rangle$, if a basis is chosen so that $|\psi\rangle$ is a member of the basis (this can always be done), $\hat{\rho}$ is a matrix with every element zero except the one diagonal element corresponding to $|\psi\rangle\langle\psi|$, which will be unity. Obviously, $\hat{\rho}^2 = \hat{\rho}$. This is less obvious in a general basis, where will not necessarily be diagonal. But the statement remains true under a transformation to a new basis.

For a mixed state, let's say for example a mixture of orthogonal states $|\psi_1\rangle, |\psi_2\rangle$, if we choose a basis including both states, the density matrix will be diagonal with just two entries w_1, w_2 Both these numbers must be less than unity, so $\hat{\rho}^2 \neq \hat{\rho}$. A mix of nonorthogonal states is left as an exercise for the reader.

SOME SIMPLE EXAMPLES

First, our case *A* above (pure state): all spins in state .

$$|\uparrow_x\rangle = (1/\sqrt{2})(|\uparrow\rangle + |\downarrow\rangle)$$

In the standard $|\uparrow\rangle, |\downarrow\rangle$ basis,

$$\hat{\rho} = |\uparrow_x\rangle\langle\uparrow_x| = \begin{pmatrix} 1/\sqrt{2} \\ 1/\sqrt{2} \end{pmatrix} \begin{pmatrix} 1/\sqrt{2} & 1/\sqrt{2} \end{pmatrix} = \begin{pmatrix} 1/2 & 1/2 \\ 1/2 & 1/2 \end{pmatrix}$$

and

$$\langle s_x \rangle = \mathrm{Tr}(\hat{\rho} s_x) = \frac{\hbar}{2}\mathrm{Tr}\begin{pmatrix} 1/2 & 1/2 \\ 1/2 & 1/2 \end{pmatrix}\begin{pmatrix} 0 & 1 \\ 1 & 0 \end{pmatrix} = \frac{\hbar}{2}$$

$$\langle s_z \rangle = \mathrm{Tr}(\hat{\rho} s_z) = \frac{\hbar}{2}\mathrm{Tr}\begin{pmatrix} 1/2 & 1/2 \\ 1/2 & 1/2 \end{pmatrix}\begin{pmatrix} 1 & 0 \\ 0 & -1 \end{pmatrix} = 0.$$

Notice that $\hat{\rho}^2 = \hat{\rho}$.

Now, case *B* (50-50 mixed up and down):

50per cent in the state $|\uparrow\rangle$, 50per cent $|\downarrow\rangle$.

The density matrix is

$$\hat{\rho} = \frac{1}{2}|\uparrow\rangle\langle\uparrow| + \frac{1}{2}|\downarrow\rangle\langle\downarrow|$$

$$= \frac{1}{2}\begin{pmatrix} 1 \\ 0 \end{pmatrix}\begin{pmatrix} 1 & 0 \end{pmatrix} + \frac{1}{2}\begin{pmatrix} 0 \\ 1 \end{pmatrix}\begin{pmatrix} 0 & 1 \end{pmatrix} = \frac{1}{2}\begin{pmatrix} 1 & 0 \\ 0 & 1 \end{pmatrix}.$$

This is proportional to the unit matrix, so

$$\mathrm{Tr}\,\hat{\rho} s_x = \frac{1}{2}\frac{\hbar}{2}\mathrm{Tr}\,\sigma_x = 0,$$

and similarly for s_y and s_z, since the Pauli *s*-matrices are all traceless. Note also that $\hat{\rho}^2 = \frac{1}{2}\hat{\rho} \neq \hat{\rho}$, as is true for all mixed states.

Finally, a 50-50 mixed state relative to the x-axis:

That is, 50per cent of the spins in the state $|\uparrow_x\rangle = (1/\sqrt{2})(|\uparrow\rangle + |\downarrow\rangle)$, "up" along the x-axis, and 50per cent in $|\downarrow_x\rangle = (1/\sqrt{2})(|\uparrow\rangle - |\downarrow\rangle)$, "down" in the x-direction.

It is easy to check that

$$\hat{\rho} = \frac{1}{2}|\uparrow_x\rangle\langle\uparrow_x| + \frac{1}{2}|\downarrow_x\rangle\langle\downarrow_x| = \frac{1}{2}\begin{pmatrix} 1/2 & 1/2 \\ 1/2 & 1/2 \end{pmatrix} + \frac{1}{2}\begin{pmatrix} 1/2 & -1/2 \\ -1/2 & 1/2 \end{pmatrix} = \frac{1}{2}\begin{pmatrix} 1 & 0 \\ 0 & 1 \end{pmatrix}$$

This is exactly the same density matrix we found for 50per cent in the state $|\uparrow\rangle$, 50per cent $|\downarrow\rangle$!

The reason is that both formulations describe a state about which we know nothing we are in a state of *total ignorance*, the spins are completely random, all directions are equally likely. The density matrix describing such a state cannot depend on the direction we choose for our axes.

Another two-state quantum system that can be analysed in the same way is the polarization state of a beam of light, the basis states being polarization in the x-direction and polarization in the y-direction, for a beam traveling parallel to the z-axis. Ordinary unpolarized light corresponds to the random mixed state, with the same density matrix as in the last example above.

TIME EVOLUTION OF THE DENSITY MATRIX

In the mixed state, the quantum states evolve independently according to Schrödinger's equation, so

$$i\hbar\frac{d\hat{\rho}}{dt} = \sum w_i H|\psi_i\rangle\langle\psi_i| - \sum w_i|\psi_i\rangle\langle\psi_i|H = [H, \hat{\rho}].$$

Note that this has the opposite sign from the evolution of a Heisenberg operator, not surprising since the density operator is made up of Schrödinger bras and kets.

The equation is the quantum analogue of *Liouville's theorem* in statistical mechanics. Liouville's theorem describes the evolution in time of an ensemble of identical classical systems, such as many boxes each filled with the same amount of the same gas at the same temperature, but the positions and momenta of the individual atoms are randomly different in each. Each box can be classically described by a single point in a huge dimensional space, a space having six dimensions for each atom (position and momentum, we ignore possible internal degrees of freedom). The whole ensemble, then, is a gas of these points in this huge space, and the rate of change of local density of this gas, from Hamilton's equations, is $\partial\rho/\partial t = -\{\rho, H\}$, the bracket now being a Poisson bracket. Anyway, this is the classical precursor of, and the reason for the name of, the density matrix.

THERMAL EQUILIBRIUM

A system in thermal equilibrium is represented in statistical mechanics by a *canonical ensemble*. If the eigenstate $|i\rangle$ of the Hamiltonian has energy E_i, the relative probability of the system being in that state is $e^{-E_i/kT} = e^{-\beta E_i}$ in the standard notation. Therefore the density matrix is:

$$\hat{\rho} = \frac{1}{Z}\sum_i e^{-\beta E_i}\,|i\rangle\langle i| = \frac{e^{-\beta H}}{Z},$$

where

$$Z = \sum_i e^{-\beta E_i} = \mathrm{Tr}\, e^{-\beta H}$$

Notice that in this formulation, apart from the normalization constant Z, the density operator is analogous to the propagator $U(t) = e^{-iHt/\hbar}$ for an imaginary time $t = -i\hbar\beta$. Incidentally, for interacting quantum fields, the propagator can be constructed as a set of Feynman diagrams corresponding to all possible sequences of particle scatterings by interaction. To find the thermodynamic properties of a field theory at finite temperature, essentially the same set of diagrams is used to find the free energy: the diagrams now describe the system propagating for a finite imaginary time, the same mathematical tools can be used.

At zero temperature ($\beta = \infty$) the probability coefficients $w_i = e^{-\beta E_i / z}$ are all zero except for the ground state: the system is in a pure state, and the density matrix has every element zero except for a single element on the diagonal. At infinite temperature, all the w_i are equal: the density matrix is just $1/N$ times the unit matrix, where N is the total number of states available to the system. In fact, the *entropy* of the system can be expressed in terms of the density matrix: $S = -kTr(\hat{\rho} In \hat{p})$. This is not as bad as it looks: both operators are diagonal in the energy subspace.

Chapter 21

Charged Particle in a Magnetic Field

INTRODUCTION

Classically, the force on a charged particle in electric and magnetic fields is given by the Lorentz force law:

$$\vec{F} = q\left(\vec{E} + \frac{\vec{v} \times \vec{B}}{c}\right)$$

This velocity-dependent force is quite different from the conservative forces from potentials that we have dealt with so far, and the recipe for going from classical to quantum mechanics replacing momenta with the appropriate derivative operators has to be carried out with more care. We begin by demonstrating how the Lorentz force law arises classically in the Lagrangian and Hamiltonian formulations.

LAWS OF CLASSICAL MECHANICS

That the Principle of Least Action leads to the Euler-Lagrange equations for the Lagrangian L:

$$\frac{d}{dt}\left(\frac{\partial L(q_i, \dot{q}_i)}{\partial \dot{q}_i}\right) - \frac{\partial L(q_i, \dot{q}_i)}{\partial q_i} = 0, \; q_i, \dot{q}_i \text{ being coordinates and velocities.}$$

The canonical momentum p_i is defined by the equation

$$p_i = \frac{\partial L}{\partial \dot{q}_i}$$

and the Hamiltonian is defined by performing a Legendre transformation of the Lagrangian:

$$H(q_i, p_i) = \Sigma p_i q_i - L(q_i, \overline{q}_i) - L(q_i, \overline{q}_i)$$

It is straightforward to check that the equations of motion can be written:

$$\overline{q}_i = \frac{\partial H}{\partial p_i}, \overline{p}_i = -\frac{\partial H}{\partial q_i}$$

These are known as *Hamilton's Equations.* Note that if the Hamiltonian is independent of a particular coordinate q_i, the corresponding momentum p_i remains constant. (Such a coordinate is termed *cyclic,* because the most common example is an angular coordinate in a spherically symmetric Hamiltonian, where angular momentum remains constant.)

For the conservative forces we have been considering so far, $L = T - V$, $H = T + V$, with T the kinetic energy, V the potential energy.

POISSON BRACKETS

Any dynamical variable f in the system is some function of the q_i's and p_i's and (assuming it does not depend explicitly on time) its development is given by:

$$\frac{d}{dt} f(q_i, p_i) = \frac{\partial f}{\partial q_i}\dot{q}_i + \frac{\partial f}{\partial p_i}\dot{p}_i = \frac{\partial f}{\partial q_i}\frac{\partial H}{\partial p_i} - \frac{\partial f}{\partial p_i}\frac{\partial H}{\partial q_i} = \{f, H\}.$$

The curly brackets are called *Poisson Brackets,* and are defined for any dynamical variables as:

$$\{A, B\} = \frac{\partial A}{\partial q_i}\frac{\partial B}{\partial p_i} - \frac{\partial A}{\partial p_i}\frac{\partial B}{\partial q_i}.$$

We have shown from Hamilton's equations that for any variable $\overline{f} = (f, H)$.

It is easy to check that for the coordinates and canonical momenta,

$$\{q_i, q_j\} = 0 = \{p_i, p_j\}, \quad \{q_i, p_j\} = \delta_{ij}.$$

This was the classical mathematical structure that led Dirac to link up classical and quantum mechanics: he realised that the Poisson brackets were the classical version of the commutators, so a classical canonical momentum must correspond to the quantum differential operator in the corresponding coordinate.

PARTICLE IN A MAGNETIC FIELD

The Lorentz force is velocity dependent, so cannot be just the gradient of some potential. Nevertheless, the classical particle path is still given by the Principle of Least Action. The electric and magnetic fields can be written in terms of a scalar and a vector potential:

$$\vec{B} = \vec{\nabla} \times \vec{A}, \quad \vec{E} = -\vec{\nabla}\varphi - \frac{1}{c}\frac{\partial \vec{A}}{\partial t}.$$

The right Lagrangian turns out to be:

$$L = \tfrac{1}{2} m\vec{v}^2 - q\varphi + \frac{q}{c}\vec{v}.\vec{A}$$

(*Note*: if you're familiar with Relativity, the interaction term here looks less arbitrary: the relativistic version would have the relativistically invariant $(q/c)\int A^{\mu} dx_{\mu}$ added to the action integral, where the four-potential $A_{\mu} = (A, \varphi)$ and $dx_{\mu} = (dx_1, dx_2, dx_3, cdt)$. This is the simplest possible invariant interaction between the electromagnetic field and the particle's four-velocity. Then in the nonrelativistic limit, $(q/c)\int A^{\mu} dx_{\mu}$ just becomes $\int q(\vec{v}.\vec{A}/c - \varphi)dt$.)

Note that for zero vector potential, the Lagrangian has the usual T - V form.

For this one-particle problem, the general coordinates q_i are just the Cartesian co-ordinates $x_i = (x_1, x_2, x_3)$, the position of the particle, and the are the three components $\dot{x}_i = v_i$ of the particle's velocity.

The important *new* point is that the canonical momentum

$$p_i = \frac{\partial L}{\partial \dot{q}_i} = \frac{\partial L}{\partial \dot{x}_i} = mv_i + \frac{q}{c} A_i$$

is no longer mass × velocity there is an extra term!

The Hamiltonian is

$$\begin{aligned} H(q_i, p_i) &= \sum p_i \dot{q}_i - L(q_i, \dot{q}_i) \\ &= \sum \left(mv_i + \frac{q}{c} A_i \right) v_i - \tfrac{1}{2} m\vec{v}^2 + q\varphi - \frac{q}{c} \vec{v}.\vec{A} \\ &= \tfrac{1}{2} m\vec{v}^2 + q\varphi \end{aligned}$$

Reassuringly, the Hamiltonian just has the familiar form of kinetic energy plus potential energy. However, to get Hamilton's equations of motion, the Hamiltonian has to be expressed solely in terms of the coordinates and canonical momenta. That is,

$$H = \frac{\left(\vec{p} - q\vec{A}(\vec{x},t)/c\right)^2}{2m} + q\varphi(\vec{x},t)$$

where we have noted explicitly that the potentials mean those at the position $\vec{x}$ of the particle at time t.

Let us now consider Hamilton's equations

$$\dot{x}_i = \frac{\partial H}{\partial p_i}, \quad \dot{p}_i = -\frac{\partial H}{\partial x_i}$$

It is easy to see how the first equation comes out, bearing in mind that

$$p_i = mv_i + \frac{q}{c} A_i = m\dot{x}_i + \frac{q}{c} A_i.$$

The second equation yields the Lorentz force law, but is a little more tricky. The first point to bear in mind is that dp/dt is *not* the acceleration, the A term also varies in time, and in a quite complicated way, since it is *the field at a point moving with the particle.* That is,

$$\dot{p}_i = m\ddot{x}_i + \frac{q}{c}\dot{A}_i = m\ddot{x}_i + \frac{q}{c}\left(\frac{\partial A_i}{\partial t} + v_j \nabla_j A_i\right).$$

The right-hand side of the second Hamilton equation $\bar{p}_i = -\frac{\partial H}{\partial x_i}$ is

$$-\frac{\partial H}{\partial x_i} = \frac{\left(\vec{p} - q\vec{A}(\vec{x},t)/c\right)}{m}\cdot\frac{q}{c}\cdot\frac{\partial \vec{A}}{\partial x_i} - q\frac{\partial \varphi(\vec{x},t)}{\partial x_i}$$

$$= \frac{q}{c}v_j \nabla_i A_j - q\nabla_i \varphi$$

Putting the two sides together, the Hamilton equation reads:

$$m\ddot{x}_i = -\frac{q}{c}\left(\frac{\partial A_i}{\partial t} + v_j \nabla_j A_i\right) + \frac{q}{c}v_j \nabla_i A_j - q\nabla_i \varphi.$$

Using $\vec{v}\times(\vec{\nabla}\times\vec{A}) = \vec{\nabla}(\vec{v}.\vec{A}) - (\vec{v}.\vec{\nabla})\vec{A}$, $\vec{B} = \vec{\nabla}\times\vec{A}$, and the expressions for the electric and magnetic fields in terms of the potentials, the Lorentz force law emerges:

$$m\ddot{\vec{x}} = q\left(\vec{E} + \frac{\vec{v}\times\vec{B}}{c}\right)$$

QUANTUM MECHANICS OF A PARTICLE IN A MAGNETIC FIELD

We make the standard substitution:

$\vec{p} = -i\hbar\vec{\nabla}$, so that $[x_i, p_j] = i\hbar\delta_{ij}$ as usual: but now $p_i \neq mv_i$.

This leads to the novel situation that the velocities in different directions *do not commute*. From

$$mv_i = -i\hbar\nabla_i - qA_i/c$$

it is easy to check that

$$[v_x, v_y] = \frac{iq\hbar}{m^2 c}B$$

To actually solve Schrödinger's equation for an electron confined to a plane in a uniform perpendicular magnetic field, it is convenient to use the Landau gauge,

$$\vec{A}(x,y,z) = (-By,0,0)$$

giving a constant field B in the z direction. The equation is

$$H\psi(x,y) = \left[\frac{1}{2m}(p_x + qBy/c)^2 + \frac{p_y^2}{2m}\right]\psi(x,y) = E\psi(x,y).$$

Note that x *does not appear in this Hamiltonian,* so it is a cyclic coordinate, and p_x is conserved. In other words, this H commutes with p_x, so H and p_x have a common set of eigenstates. We know the eigenstates of p_x are just the plane waves $e^{ip_x x/\hbar}$, so the common eigenstates must have the form:

$$\psi(x,y) = e^{ip_x x/\hbar}\chi(y)$$

Operating on this wavefunction with the Hamiltonian, the operator p_x appearing in H simply gives its eigenvalue. That is, the p_x in H just becomes a number! Therefore, writing $p_y = -i\hbar d/dy$, the y-component $\chi(y)$ of the wavefunction satisfies:

$$-\frac{\hbar^2}{2m}\frac{d^2}{dy^2}\chi(y) + \tfrac{1}{2}m\left(\frac{qB}{mc}\right)^2 (y-y_0)^2\chi(y) = E\chi(y)$$

where

$$y_0 = -cp_x/qB$$

We now see that the conserved canonical momentum p_x in the x-direction is actually the coordinate of the center of a simple harmonic oscillator potential in the y-direction! This simple harmonic oscillator has frequency $\omega = |q|B/mc$, so the allowed values of energy for a particle in a plane in a perpendicular magnetic field are:

$$E = (n+\tfrac{1}{2})\hbar\omega = (n+\tfrac{1}{2})\hbar|q|B/mc.$$

The frequency is of course the cyclotron frequency that of the classical electron in a circular orbit in the field.

Let us confine our attention to states corresponding to the lowest oscillator state, . How many such states are there? Consider a square of conductor, area $A = L_x \times L_y$, and, for simplicity, take periodic boundary conditions. The center of the oscillator wave function y_0 must lie between 0 and L_y. But remember that $y_0 = -cp_x / qB$, and with periodic boundary conditions $e^{ip_x L_x/\hbar} = 1$, so $p_x = 2n\pi\hbar / L_x = nh / L_x$. This means that y_0 takes a series of evenly-spaced discrete values, separated by

$$\Delta y_0 = ch / qBL_x.$$

So the total number of states $N = L_y / \Delta y_0$,

$$N = \frac{L_x L_y}{\left(\frac{hc}{qB}\right)} = A.\frac{B}{\Phi_0},$$

where ϕ_0 is called the "flux quantum". So the total number of states in the lowest energy level $E = \frac{1}{2}\hbar\omega$ (usually referred to as the lowest Landau level, or *LLL*) is exactly equal to the total number of flux quanta making up the field *B* penetrating the area *A*.

It is instructive to find y_0 from a purely classical analysis.

Writing $m\dot{\vec{v}} = \frac{q}{c}\vec{v} \times \vec{B}$ in components,

$$m\ddot{x} = \frac{qB}{c}\dot{y},$$

$$m\ddot{y} = -\frac{qB}{c}\dot{x}.$$

These equations integrate trivially to give:

$$m\dot{x} = \frac{qB}{c}(y - y_0),$$

$$m\dot{y} = -\frac{qB}{c}(x - x_0).$$

Here (x_0, y_0) are the coordinates of the center of the classical circular motion (the velocity vector $\dot{\vec{r}} = (\dot{x}, \dot{y})$ is always perpendicular to $(\vec{r} - \vec{r}_0)$), and $\vec{r}_0$ is given by

$$y_0 = y - cmv_x / qB = -cp_x / qB$$
$$x_0 = x + cmv_y / qB = x + cp_y / qB.$$

(Recall that we are using the gauge $\vec{A}(x, y, z) = (-By, 0, 0)$, and $p_x = \dfrac{\partial L}{\partial \dot{x}} = mv_x + \dfrac{q}{c} A_x$, etc.)

Chapter 22

Tensor Operators

REVIEW OF THE ROTATION OPERATOR

Recall that the matrix elements of the rotation operator

$$U\left(R\left(\vec{\theta}\right)\right)=e^{-\frac{i\vec{\theta}\cdot\vec{J}}{\hbar}}$$

within a definite j subspace are written

$$D^{(j)}_{m'm}\left(R\left(\vec{\theta}\right)\right)=\langle j,m'|e^{-\frac{i\vec{\theta}\cdot\vec{J}}{\hbar}}|j,m\rangle.$$

ROTATING A KET

By a ket, we mean here an element of the j subspace expressed as a sum over the basis kets $|j, m\rangle$ (which we'll just write as $|m\rangle$ here) with a set of coefficients $a_{m'}$,

$$|\alpha\rangle=\sum_{m=-j}^{j}\alpha_m|m\rangle$$

and we are finding how that *set of coefficients* transforms under rotation. (Any such ket will stay within the j subspace under rotation, because the rotation operator commutes with $\vec{j}^2$.)

Rotating the ket by $\vec{\theta}$ to $|\alpha'\rangle=\sum\alpha'_{m'}|m'\rangle$,

$$|\alpha'\rangle=e^{-\frac{i\vec{\theta}\cdot\vec{J}}{\hbar}}|\alpha\rangle$$

$$=\sum_{m',m}\alpha_m|m'\rangle\langle m'|e^{-\frac{i\vec{\theta}\cdot\vec{J}}{\hbar}}|m\rangle$$

$$= \sum_{m',m} D^{(j)}_{m'm} \alpha_m |m'\rangle$$

So the ket rotation transformation is

$$\alpha'_{m'} = \sum_{m',m} D^{(j)}_{m'm} \alpha_{m'}, \quad \text{or } \alpha' = D\alpha$$

with the usual matrix-multiplication rules.

Rotating a Basis Vector $|j,m>$

Now we look at something slightly different: suppose we apply the rotation operator to one of the *basis vectors* $|j,m>$, what is the result?

$$e^{-\frac{i\vec{\varphi}\cdot\vec{J}}{\hbar}}|j,m\rangle = \sum_{m'}|j,m'\rangle\langle j,m'|e^{-\frac{i\vec{\varphi}\cdot\vec{J}}{\hbar}}|j,m\rangle = \sum_{m'}|j,m'\rangle D^{(j)}_{m'm}(R).$$

Notice the reversal of m, m' compared with the operation on the set of component coefficients of the ket!

ROTATING AN OPERATOR

Just as in the Schrödinger versus Heisenberg formulations, we can either apply the rotation operator to the kets and leave the operators alone, or we can leave the kets alone, and rotate the operators:

$$A \to e^{\frac{i\vec{\varphi}\cdot\vec{J}}{\hbar}} A e^{-\frac{i\vec{\varphi}\cdot\vec{J}}{\hbar}} = U^\dagger A U$$

which will yield the same matrix elements, so the same physics.

Confusingly, there is a slightly different situation in which we need to rotate an operator, and it gives an opposite result. Suppose an operator T acts on a ket $|\alpha>$ to give the ket $T|\alpha>$. For kets $|\alpha>$ and $T|\alpha>$ to go to $U|\alpha>$ and $UT|\alpha>$ respectively under a rotation U, T itself must transform as $T \to UTU^\dagger$ (recall $U^\dagger = U^{-1}$).

SCALAR OPERATORS

A scalar operator is an operator which is invariant under rotations, for example the Hamiltonian of a particle in a spherically symmetric potential. (There are many less trivial examples of scalar operators, such as the dot product of two vector operators, as in a spin-orbit coupling.)

The transformation of an operator under an infinitesimal rotation is given by:

$$S \to U^{\dagger}(R) S U(R) \text{ with } U(R) = 1 - \frac{i\varepsilon \mathbf{J}\cdot\hat{\mathbf{n}}}{\hbar}$$

from which

$$S \to S + \left[\frac{i\varepsilon \mathbf{J}\cdot\hat{\mathbf{n}}}{\hbar}, S\right].$$

It follows that a scalar operator S, which does not change at all, must commute with all the components of the angular momentum operator, and hence must have a common set of eigenkets with, say, J^2 and J_z.

VECTOR OPERATORS

We have of course been using vector operators all along, for example $\vec{x}, \vec{L}, \vec{S}$ and so on. A vector operator has three components, corresponding to the Cartesian x, y, and z axes. Classically, such a three-dimensional vector transforms under rotation as

$$V_i \to \sum R_{ij} V_j,$$

with the usual rotation matrix, for example

$$R_z(\theta) = \begin{pmatrix} \cos\theta & -\sin\theta & 0 \\ \sin\theta & \cos\theta & 0 \\ 0 & 0 & 1 \end{pmatrix}$$

for rotation about the z-axis.

A quantum mechanical vector operator $\vec{V}$ is *defined* by requiring that the expectation values of its three components in any state transform like the components of a classical vector under rotation. It follows from this that the operator itself must transform vectorially,

$$V_i' = U^{\dagger}(R) V_i U(R) = \sum R_{ij} V_j$$

To see what this implies, it is easiest to look at a simple case. For an infinitesimal rotation about the z-axis,

$$R_z(\varepsilon) = \begin{pmatrix} 1 & -\varepsilon & 0 \\ \varepsilon & 1 & 0 \\ 0 & 0 & 1 \end{pmatrix}$$

the vector transforms

$$\begin{pmatrix} V_x \\ V_y \\ V_z \end{pmatrix} \to \begin{pmatrix} 1 & -\varepsilon & 0 \\ \varepsilon & 1 & 0 \\ 0 & 0 & 1 \end{pmatrix}\begin{pmatrix} V_x \\ V_y \\ V_z \end{pmatrix} = \begin{pmatrix} V_x - \varepsilon V_y \\ V_y + \varepsilon V_x \\ V_z \end{pmatrix}$$

The unitary Hilbert space operator U corresponding to this rotation $U(R_z(\varepsilon)) = 1 - \frac{i\varepsilon J_z}{\hbar}$, so

$$U^{\dagger} V_i U = (1 + i\varepsilon J_z / \hbar) V_i (1 - i\varepsilon J_z / \hbar)$$

$$= V_i + \frac{i\varepsilon}{\hbar}[J_z, V_i].$$

The requirement that the two transformations above, the infinitesimal classical rotation generated by $R_z(\varepsilon)$ and the infinitesimal unitary transformation $U^{\dagger}(R) V_i U(R)$, are in fact the same thing yields the commutation relations of a vector operator with angular momentum:

$$i[J_z, V_x] = -\hbar V_y$$

$$i[J_z, V_y] = +\hbar V_x.$$

From this result and its cyclic equivalents, the components of *any* vector operator $\vec{V}$ must satisfy:

$$[V_i, J_j] = i\varepsilon_{ijk}\hbar V_k.$$

CARTESIAN TENSOR OPERATORS

The standard definition of a *Cartesian tensor* is that each of its suffixes transforms under rotation as do the components of an ordinary three-dimensional vector. Specifically, the nine components of a second rank tensor transform as:

$$T_{ij} \to T'_{ij} = \sum\sum R_{ii'} R_{jj'} T_{i'j'}.$$

A particular example of a second-rank tensor is $T_{ij} = U_i V_j$, where $\vec{U}$ and $\vec{V}$ are ordinary three-dimensional vectors.

The problem with this tensor is that it is *reducible* combinations of the elements can be arranged in groups such that rotations operate only within these groups. This is made evident by writing:

$$U_i V_j = \frac{\vec{U}\cdot\vec{V}}{3}\delta_{ij} + \frac{(U_i V_j - U_j V_i)}{2} + \left(\frac{U_i V_j + U_j V_i}{2} - \frac{\vec{U}\cdot\vec{V}}{3}\delta_{ij}\right).$$

The first term, the dot product of the two vectors, is clearly a *scalar* under rotation, the second term, which is an antisymmetric tensor has three independent components which are the *vector* components of the vector product $\vec{U} \times \vec{V}$, and the third term is a *symmetric traceless tensor*, which has five independent components. Altogether, then, there are 1 + 3 + 5 = 9 components, as required.

SPHERICAL TENSORS

Notice the numbers of elements of these irreducible subgroups: 1, 3, 5. These are exactly the numbers of elements of angular momenta representations for j = 0, 1, 2! The first term is of course trivial: the scalar by definition is not affected by rotation, and neither is an j = 0 state.

To deal with the second and third terms, we introduce tensor operators having three and five components, such that under rotation these sets of components transform among themselves just as do the sets of eigenkets of angular momentum in the j = 1 and j = 2 representations respectively.

Recall that the matrix elements of the rotation operator

$$U\left(R\left(\vec{\theta}\right)\right) = e^{-\frac{i\vec{\theta}\cdot\vec{J}}{\hbar}}$$

within a definite j subspace are written

$$D^{j}_{m'm}\left(R\left(\vec{\theta}\right)\right) = \langle j,m'|e^{-\frac{i\vec{\theta}\cdot\vec{J}}{\hbar}}|j,m\rangle$$

so under rotation operator a basis state $| j,m >$ transforms as:

$$e^{-\frac{i\vec{\theta}.\vec{J}}{\hbar}}|j,m\rangle = \sum_{m'}|j,m'\rangle\langle j,m'|e^{-\frac{i\vec{\theta}.\vec{J}}{\hbar}}|j,m\rangle = \sum_{m'}|j,m'\rangle D^{(j)}_{m'm}(R).$$

Notice the reversal of m, m' compared with the usual operation of a matrix on a vector: this is because we are rotating the basis here, rather than rotating relative to the basis.

The basic idea is that these irreducible subgroups into which Cartesian tensors apparently decompose under rotation (generalizing from our one example) form a more natural basis set of tensors for problems with rotational symmetries.

Definition: We define a spherical tensor of rank k as a set of $2k + 1$ components T_k^q, $q = k, k-1, \ldots, -k$ such that under rotation they transform like a set of angular momentum eigenkets, that is,

$$U(R)T_k^q U^\dagger(R) = \sum_{q'} D^{(k)}_{q'q} T_k^{q'}.$$

To see the properties of these spherical tensors, it is useful to evaluate the above equation for infinitesimal rotations, for which , $D^{(k)}_{q,q}(\vec{\varepsilon}) = \langle k,q'| I - i\vec{\varepsilon}.\vec{j}/\hbar | k,q\rangle$ to find

$$\left[J_\pm, T_k^q\right] = \pm\hbar\sqrt{(k\mp q)(k\pm q+1)}\, T_k^{q\pm 1}$$

$$\left[J_z, T_k^q\right] = \hbar q T_k^q.$$

This set of commutation relations could be taken as the *definition* of the spherical tensors.

A SPHERICAL VECTOR

The $j = 1$ angular momentum eigenkets are just the familiar spherical harmonics

$$Y_1^0 = \sqrt{\frac{3}{4\pi}}\frac{z}{r}, \quad Y_1^{\pm 1} = \mp\sqrt{\frac{3}{4\pi}}\frac{x \pm iy}{\sqrt{2}r}.$$

The rotation operator will transform (x, y, z) as an ordinary vector in three-space, and this is evidently equivalent to

$$|j=1,m\rangle \to \sum_{m'} |j=1,m'\rangle D^{(j)}_{m'm}(R)$$

It follows that the spherical representation of a three vector (v_x, v_y, v_z) has the form:

$$T_1^{\pm 1} = \mp \frac{V_x \pm i V_y}{\sqrt{2}} = V_1^{\pm 1}, \quad T_1^0 = V_z = V_1^0.$$

In line with spherical tensor notation, the components $\left(T_1^1, T_1^0, T_1^{-1}\right)$ are denoted T_1^a.

THE WIGNER-ECKART THEOREM

At this point, we must bear in mind that these tensor operators are not necessarily just functions of angle. For example, the position operator is a spherical vector multiplied by the radial variable *r*, and kets specifying atomic eigenstates will include radial quantum numbers as well as angular momentum.

Therefore, the matrix element of a tensor between two states will look like $\left\langle a_2, j_2, m_2 \mid T_k^q \mid a_1, j_1, m_1 \right\rangle$, where the *j*'s and *m*'s denote the usual angular momentum eigenstates and the *a*'s are nonangular quantum numbers, such as radial states.

The basic point of the Wigner–Eckart theorem is that the angular dependence of these matrix elements can be factored out, and that it is given by the Clebsch-Gordan coefficients. Having factored it out, the remaining dependence, which is only on the total angular momentum in each of the kets, not the relative orientation, and of course on the a's, is traditionally written as a bracket with double lines, that is,

$$\langle \alpha_2, j_2, m_2 | T_k^q | \alpha_1, j_1, m_1 \rangle = \langle \alpha_2, j_2 \| T_k \| \alpha_1, j_1 \rangle \langle j_2, m_2 | k, q; j_1, m_1 \rangle.$$

for example, and is not that difficult. The basic strategy is to put the defining identities

$$\left[J_{\pm}, T_k^q\right] = \pm\hbar\sqrt{(k \mp q)(k \pm q + 1)}\, T_k^{q\pm1}$$
$$\left[J_z, T_k^q\right] = \hbar q T_k^q$$

between $|a,k,m>$ bras and kets, then get rid of the $J_{\pm}$ and J_z by having them operate on the bra or ket. This generates a series of linear equations for $\left\langle \alpha_2, j_2, m_2 \,|\, T_k^q \,|\, \alpha_1, j_1, m_1 \right\rangle$ matrix elements with m variables differing by one, and in fact this set of linear equations is *identical* to the set that generates the Clebsch-Gordan coefficients, so we must conclude that these spherical tensor matrix elements, ranging over possible m and j values, are exactly proportional to the Clebsch-Gordan coefficients and that is the theorem.

A Few Hints for 15.3.3: that first matrix element comes from adding a spin j to a spin 1, writing the usual maximum m state, applying the lowering operator to both sides to get the total angular momentum $j + 1$, $m = j$ state, then finding the same m state orthogonal to that, which corresponds to total angular momentum j (instead of $j + 1$).

For the operator J, the Wigner-Eckart matrix element simplifies because J cannot affect a, and also it commutes with J^2, so cannot change the total angular momentum.

So, in the Wigner-Eckart equation, replace T_k^q on the left-hand side by J_1^0, which is just J_z. The result of (1) should follow.

(2) First note that a scalar operator cannot change m. Since c is independent of A we can take $A = J$ to find c.

Chapter 23

Variational Methods

VARIATIONAL METHOD FOR FINDING THE GROUND STATE ENERGY

The idea is to guess the ground state wave function, but the guess must have an adjustable parameter, which can then be varied (hence the name) to minimize the expectation value of the energy, and thereby find the best approximation to the true ground state wave function. This crude sounding approach can in fact give a surprisingly good approximation to the ground state energy (but usually not so good for the wave function, as will become clear).

We'll begin with a single particle in a potential, $H = p^2/2m + V(\vec{r})$. If the particle is restricted to one dimension, and we're looking for the ground state in any fairly localized potential well, we can start with the family of normalized Gaussians, $|\psi,\alpha> = \left(\frac{\alpha}{\pi}\right)^{1/4} e^{-\alpha x^2/2}$: just find $\langle\psi,\alpha|H|\psi,\alpha\rangle$, differentiate the result with respect to , setting this to zero (and checking that you have in fact found a minimum.) Not surprisingly, this gives the exact ground state for the simple harmonic oscillator potential, and for nothing else. What is perhaps surprising is that the result is only off by 30per cent or so for the attractive delta-function potential, even though the wave function looks a lot different .Obviously, the Gaussian family cannot be used if there is an infinite wall anywhere: one must find a family of wave functions vanishing at the wall.

To gain some insight into what we're doing, suppose the Hamiltonian $H = p^2/2m + v(\vec{r})$ has the set of (unknown to us) eigenstates

$$H|n>= E_n|n>.$$

Since the Hamiltonian is Hermitian, these states span the space of possible wave functions, including our variational family, so:

$$|\psi,\alpha>= \Sigma a_n(\alpha)|n>.$$

From this,

$$\frac{\langle\psi,\alpha|H|\psi,\alpha\rangle}{\langle\psi,\alpha|\psi,\alpha\rangle} = \sum |a_n|^2 E_n \geq E_0$$

for *any* $|\psi,\alpha>$. (We don't need the denominator if we've chosen a family of normalized wave functions, as we did with the Gaussians above.) Evidently, minimizing $\frac{\langle\psi,\alpha|H|\psi,\alpha\rangle}{\langle\psi,\alpha|\psi,\alpha\rangle}$ as a function of α gives us an upper bound on the ground state energy, hopefully not too far from the true value.

We can see immediately that this will probably be better for finding for the ground state energy than for mapping the ground state wave function: suppose the optimum state in our family is actually $|\alpha_{\min}\rangle = N(|0\rangle + 0.2|1\rangle)$, with the normalization constant $N \cong 0.098$, a 20per cent admixture of the first excited state. Then the wave function is off by of order 20per cent, but the energy estimate will be too high by $0.04(E_1 \ E_0)$ usually a much smaller error.

To get some idea of how well this works, Messiah applies the method to the ground state of the hydrogen atom. We know it's going to be spherically symmetric, so it amounts to a one-dimensional problem: just the radial wave function. Using standard notation,

$$a_0 = \hbar^2 / me^2,\ E_0 = me^4 / 2\hbar^2,\quad \rho = r / a_0$$

and for a trial wave function u

$$E(u) = -E_0 \frac{\int u\left(\frac{d^2}{d\rho^2} + \frac{2}{\rho}\right) u\, d\rho}{\int u^2 d\rho}$$

(we're going to take u real).

Messiah tries three families:

$$u_1 = \rho e^{-\alpha\rho}$$

$$u_2 = \frac{\rho}{\alpha^2 + \rho^2}$$

$$u_3 = \rho^2 e^{-\alpha\rho}$$

and finds $\alpha_{min} = 1, \pi/4, 3/2$ respectively. The first family, u_1, includes the exact result, and the minimization procedure finds it.

For the three families, then the energy of the best state is off by 0, 25per cent, 21per cent respectively.

The wave function error is defined as how far the square of the overlap with the true ground state wave function falls short of unity. For the three families, $\varepsilon = 1 - |\langle \psi_0 | \psi_{var} \rangle|^2$:0, 0.21, 0.05. Notice here that our hand waving argument that the energies would be found much more accurately than the wave functions comes unstuck. The third family has far better wave function overlap than the second, but only a slightly better energy estimate. Why? A key point is that the potential is singular at the origin, there is a big contribution to potential energy from a rather small region, and the third family wave function is the least accurate of the three there. The second family functions are very inaccurate at large distances: the expectation value $\langle r \rangle = 1.5\alpha_0, \infty, 1.66\alpha_0$ for the three families.

But at large distances, both kinetic and potential energies are small, so the result can still look reasonable. These examples reinforce the point that the variational method should be used cautiously.

VARIATIONAL METHOD FOR HIGHER STATES

In some cases, the approach can be used easily for higher states: specifically, in problems having some symmetry. For example, if the one dimensional attractive potential is symmetric about the origin, and has more than one bound state, the ground state will be even, the first excited state odd. Therefore, we can estimate the energy of the first excited state by minimizing a family of odd functions, such as

$$\psi(x,\alpha) = \left(\sqrt{\pi}/2\alpha^{3/2}\right) x e^{-\alpha x^2/2}$$

GROUND STATE ENERGY OF THE HELIUM ATOM BY THE VARIATIONAL METHOD

We know the ground state energy of the hydrogen atom is –1 Ryd, or –13.6 ev. The He^+ ion has Z = 2, so will have ground state energy, proportional to Z^2, equal to –4 Ryd. Therefore for the He atom, if we neglect the electron-electron interaction, the ground state energy will be -8 Ryd, –109 ev., the two electrons having opposite spins will both be in the lowest spatial state. Actually, experimentally, the He atom ground state energy is only –79 ev, because the repulsion between the electrons loosens things.

To get a better value for the ground state energy still using tractable wave functions, we change the wave functions from the ionic wave function $\left(Z^3/\pi a_0^3\right)^{1/2} e^{-Zr/a_0}$ with Z = 2 $\left(Z'^3/\pi a_0^3\right)^{1/2} e^{-Zr/a_0}$ to with Z' now a variable parameter. In other words, we are trying to allow for the electron-electron repulsion, which must push the wave functions out a bit, by keeping exactly the same shaped wave function but lessening the effective nuclear charge as reflected in the spread of the wave function from Z to Z', and we'll determine Z' by varying it to find the minimum total energy, including the term from electron-electron repulsion.

To find the potential energy from the nuclear-electron interactions, we of course use the actual nuclear charge Z = 2, but the Z′ wave function, so the nuclear *P.E.* for the two electrons is:

$$P.E. = -2Ze^2\int_0^{\infty}\frac{1}{r}4\pi r^2 dr\left(Z'^3/\pi a_0^3\right)e^{-2Z'r/a_0}$$

$$= -4ZZ'\left(e^2/2a_0\right)$$

$$= -8Z'\,\text{Ryd}\quad (Z = 2).$$

This could have been figured out from the formula for the one-electron ion, where the potential energy for the one electron is -2Z^2 Ryd, one factor of Z being from the nuclear charge, the other from the consequent shrinking of the orbit.

The kinetic energy is even easier: it depends entirely on the shape of the wave function, not on the actual nuclear charge, so for our trial wave function it has to be Z'^2 Ryds per electron.

The tricky part is the *P.E.* for the electron-electron interaction. This is positive.

Each electron has a wave function $\left(Z'^3/\pi a_0^3\right)^{1/2} e^{-Zr/a_0}$, a spherical charge probability distribution.

Denoting charge probability density by ρ(r), we need

$$I = \iint d\vec{r}_1 d\vec{r}_2 \frac{\rho(\vec{r}_1)\rho(\vec{r}_2)}{|\vec{r}_1-\vec{r}_2|}$$

$$= 16\pi^2\int_0^{\infty} r_1^2 dr_1 \int_0^{\infty} r_2^2 dr_2 \frac{\rho(r_1)\rho(r_2)}{r_>},$$

$$r_> = \max\left(r_1, r_2\right).$$

This integral must be done in two stages:

$$I = 16\pi^2 \int_0^\infty \rho(r_1) r_1 dr_1 \left(\int_0^{r_1} \rho(r_2) r_2^2 dr_2 + r_1 \int_{r_1}^\infty \rho(r_2) r_2 dr_2 \right).$$

The result is -(5/4)Z′ Collecting terms, the total energy (for Z = 2) is: $E = -2\left(4Z' - Z'^2 - \frac{5}{8}Z'\right)$ Ryd

and this is minimized by taking $Z' = 2 - \frac{5}{16}$, giving an energy of –77.5 ev, off the true value by about 1 ev, so indeed the presence of the other electron is taken care of as far as total energy is concerned by shielding the nuclear charge by an amount (5/16)*e*.

Chapter 24

The WKB Approximation

THE SEMICLASSICAL APPROXIMATION

Consider a particle moving along in a slowly varying one-dimensional potential. By "slowly varying" we mean here that in any small region the wave function is well approximated by a plane wave, and that the wavelength only changes over distances long compared with a wavelength. Under these conditions, it's easy to see the general form of the solution to the time independent Schrödinger equation

$$-\frac{\hbar^2}{2m}\frac{d^2\psi(x)}{dx^2}+V(x)\psi(x)=E\psi(x)$$

Very approximately, $\psi(x)$ will look like $A(x)e^{ip(x)x/\hbar}$ where $p(x)$ is the "local momentum" we define classically by

$$p(x)^2/2m+V(x)=E$$

and $A(x)$ is slowly varying compared with the phase factor.

Clearly this is a *semiclassical* limit: $\hbar$ has to be sufficiently small that there are many oscillations over the typical distance over which the potential varies.

To handle this a little more precisely, we emphasize the rapid phase variation of the wave function by casting it in the form

$$\psi(x) = e^{(i/\hbar)\sigma(x)}$$

From this,

$$i\hbar\psi'(x) = -\sigma'(x)e^{(i/\hbar)\sigma(x)},$$

and

$$-\hbar^2\psi''(x) = -i\hbar\sigma''(x)e^{(i/\hbar)\sigma(x)} + (\sigma'(x))^2 e^{(i/\hbar)\sigma(x)}.$$

In terms of $\sigma(x)$, therefore, the Schrödinger equation becomes:

$$-i\hbar\sigma''(x) + (\sigma'(x))^2 = (p(x))^2.$$

And, since we're assuming the system is close to classical, it makes sense to expand σ as a series in $\hbar$ (following Landau and Lifshitz):

$$\sigma = \sigma_0 + (\hbar/i)\sigma_1 + (\hbar/i)^2\sigma_2 + \ldots$$

The zeroth order approximation is

$$(\sigma_0')^2 = p^2$$

and if we fix the sign of p by

$$p(x) = +\sqrt{2m(E - V(x))}$$

we conclude that

$$\sigma_0 = \pm\int p(x)dx.$$

(As we discussed in the lecture on path integrals, in the classical limit one path dominates, and the phase of the wave function is $(i/\hbar)$ times the classical action S along that path. In the present case, $S = -Et \pm \int pdx$, we've already factored out the Et since we're dealing here with the time-independent wave function.)

From the Schrödinger equation $-i\hbar\sigma''(x)+(\sigma'(x))^2 = (p(x))^2$, this approximate solution is only valid if we can ignore that first term. That is to say, we must have $\left|\hbar\sigma''(x)/(\sigma'(x))^2\right| << 1$, or

$$\left|\frac{d(\hbar/\sigma')}{dx}\right| \ll 1.$$

But in leading approximation $\sigma' = p$, and $p = 2\pi\hbar/\lambda$, so the condition is

$$\frac{1}{2\pi}\left|\frac{d\lambda}{dx}\right| \ll 1$$

This just means the change in wavelength over a distance of one wavelength must be small. Obviously, this cannot always be the case: if the particle is confined by an attractive potential, at the edge of the classically allowed region, that is, where $E = V(x)$, p is zero and the wavelength is infinite. The approximation is only good well away from that point, to which we shall return shortly.

First, let us evaluate the next term in the $\hbar$ expansion of the Schrödinger equation:

$$-i\hbar\sigma_0'' + 2\sigma_0'(\hbar/i)\sigma_1' = 0$$

so

$$\sigma_1' = -\sigma_0''/2\sigma_0' = -p'/2p,$$

and $\sigma_1 = -\frac{1}{2}\ln p$. This means the wave function to this order of accuracy has the form

$$\psi(x) = \frac{C_1}{\sqrt{p(x)}}e^{(i/\hbar)\int p\,dx} + \frac{C_2}{\sqrt{p(x)}}e^{-(i/\hbar)\int p\,dx}.$$

To interpret the $\sqrt{p(x)}$ factor, consider the first term, a wave moving to the right. Since p is real, the exponential has modulus unity, and the local amplitude squared is proportional to $1/p$, that is, $1/v$, where v is the velocity of the particle. This is simple to understand physically: the probability of finding the particle in any given small interval is proportional to the time it spends there, hence inversely proportional to its speed.

In the classically forbidden regions, p is pure imaginary, the wave function becomes

$$\psi(x) = \frac{C_1'}{\sqrt{|p(x)|}} e^{-(1/\hbar)\int |p| dx} + \frac{C_2'}{\sqrt{|p(x)|}} e^{(1/\hbar)\int |p| dx}.$$

Let us assume we're dealing with a one-dimensional potential, and the classically allowed region is $b \le x \le a$. (I'm just following Landau's notation here.) Clearly, in the forbidden region to the right, $x > a$, only the first term in the above equation for $\psi(x)$ appears, and for $x < b$ only the second term.

CONNECTION FORMULAS, BOUNDARY CONDITIONS AND QUANTIZATION RULES

But how do we connect the three regions together? At $x = a$, we assume the potential is well-approximated by a linear increase, $E - V(x) \cong F_0(x-a)$ (so F_0 would be the force) and then approximate the function by the known exact solution for a linear everywhere potential: the Airy function. It is known that for the Airy function, the solution having the form

$$\psi(x) = \frac{C}{2\sqrt{|p(x)|}} e^{-(1/\hbar)\int_a^x |p| dx}$$

to the right becomes

$$\psi(x) = \frac{C}{\sqrt{|p(x)|}} \cos\left((1/\hbar)\int_a^x p dx + \tfrac{1}{4}\pi\right)$$

$$= \frac{C}{\sqrt{|p(x)|}} \sin\left((1/\hbar)\int_x^a p dx + \tfrac{1}{4}\pi\right)$$

for $x < a$.

At b, the same argument gives

$$\psi(x) = \frac{C}{\sqrt{|p(x)|}} \sin\left((1/\hbar) \int_b^x p dx + \tfrac{1}{4}\pi \right).$$

For these two expressions to be consistent, we must have

$$\frac{1}{\hbar}\int_b^a p dx + \tfrac{1}{2}\pi = (n+1)\pi, \text{ or } \oint p dx = 2\pi\hbar\left(n + \tfrac{1}{2}\right).$$

where the latter integral is over a complete cycle of the classical motion.

Here n is the number of zeros of the wave function.

The time for a complete classical circuit is

$$T = 2\int_b^a dx / v = 2m\int_b^a dx / p.$$

Note that $\oint p dx$ is the area of the classical path in phase space, so we see each state has an element of phase space $2\pi\hbar$. From this, we can figure out the approximate energy splitting between levels in the quasi-classical limit: the change in the integral with energy ΔE corresponding to one level must be $2\pi\hbar$ That is,

$$\Delta E \oint (\partial p / \partial E)\, dx = 2\pi\hbar.$$

Now $(\partial E / \partial p) = v$, so $\oint (\partial p / \partial E) dx = \oint dx / v = T.$

Therefore, $\Delta E = 2\pi\hbar / T = \hbar\omega.$

This is just saying that if the particle emits one photon and drops to the next level, the frequency of the photon emitted is just the orbital frequency of the particle, a very natural conclusion in the quasi-classical limit.

THE RADIAL CASE

In the above analysis for a particle confined to one dimension, the connection formulas can be understood with a simple picture: the wave function "spills over" into the forbidden regime, and its twisting there counts as an extra $\frac{1}{4}\pi$ of phase change, so in the lowest state the total phase change in the allowed region need only be $\frac{1}{2}\pi$. In the radial case, assuming the potential is well behaved at the origin, the wave function goes to zero there. A bound state will still spill over beyond the classical turning point at r_0, say, but clearly there must be a total phase change of $\frac{3}{4}\pi$ in the allowed region for the lowest state, since there can be no spill over to negative r.

The general formula will be

$$\frac{1}{\hbar}\int_0^{r_0} p(r)\,dr = \left(n+\tfrac{3}{4}\right)\pi, \quad n = 0,1,2,\ldots,$$

the series terminating if and when the turning point reaches infinity.

Warning: actually, some potentials, including the Coulomb potential and the centrifugal barrier for $l \neq 0$, are in fact singular at r = 0. These cases require special treatment.

Chapter 25

Time-Independent Perturbation Theory

INTRODUCTION

If an atom (not necessarily in its ground state) is placed in an external electric field, the energy levels shift, and the wave functions are distorted. This is called the *Stark effect*. The new energy levels and wave functions could in principle be found by writing down a complete Hamiltonian, including the external field, and finding the eigenkets. This actually can be done in one case: the hydrogen atom, but even there, if the external field is small compared with the electric field inside the atom (which is billions of volts per meter) it is easier to compute the changes in the energy levels and wave functions with a scheme of successive corrections to the zero-field values. This method, termed *perturbation theory*, is the single most important methcd of solving problems in quantum mechanics, and is widely used in atomic physics, condensed matter and particle physics.

It should be noted that there *are* problems which cannot be solved using perturbation theory, even when the perturbation is very weak, although such problems are the exception rather than the rule. One such case is the one-dimensional problem of free particles perturbed by a localized potential of strength λ. As we found earlier in the course, switching on an arbitrarily weak attractive potential causes the free particle wave function to drop below the continuum of plane wave energies and become a localized bound state with binding energy of order λ^2. However, changing the sign of λ to give a repulsive potential there is no bound state, the lowest energy plane wave state stays at energy zero. Therefore the energy shift on switching on the perturbation cannot be represented as a power series in λ, the strength of the perturbation. This particular difficulty does not in general occur in three dimensions, where arbitrarily weak potentials do not give bound states except for certain many-body problems (like the Cooper pair problem) where the exclusion principle reduces the effective dimensionality of the available states.

THE PERTURBATION SERIES

We begin with a Hamiltonian H^0 having known eigenkets and eigenenergies:

$$H^0 \left| n^0 \right\rangle = E_n^0 \left| n^0 \right\rangle.$$

The task is to find how these eigenkets and eigenenergies change if a small term H^1 (an external field, for example) is added to the Hamiltonian, so:

$$\left(H^0 + H^1\right)\left| n \right\rangle = E_n \left| n \right\rangle.$$

That is to say, on switching on H^1,

$$\left| n^0 \right\rangle \to \left| n \right\rangle, \; E_n^0 \to E_n.$$

The basic assumption in perturbation theory is that H^1 is sufficiently small that the leading corrections are the same order of magnitude as H^1 itself, and the true energies can be better and better approximated by a successive series of corrections, each of order H^1/H^0 compared with the previous one.

The strategy, then, is to expand the true wave function and corresponding eigenenergy as series in H^1/H^0. These series are then fed into $\left(H^0 + H^1\right)| n >= E_n | n >$, and terms of the same order of magnitude in H^1/H^0 on the two sides are set equal. The equations thus generated are solved one by one to give progressively more accurate results.

To make it easier to identify terms of the same order in H^1/H^0 on the two sides of the equation, it is convenient to introduce a dimensionless parameter λ which always goes with H^1, and then expand $|n>, E_n$ as power series in λ, $|n\rangle = |n^0\rangle + \lambda|n^1\rangle + \lambda^2|n^2\rangle + \dots$, etc. The ket $|n^m>$ multiplied by λ^m is therefore of order $\left(H^1/H^0\right)^m$.

This λ is purely a bookkeeping device: we will set it equal to 1 when we are through! It's just there to keep track of the orders of magnitudes of the various terms.

Putting the series expansions for $|n>, E_n$ in

$$\left(H^0 + \lambda H^1\right)|n\rangle = E_n|n\rangle$$

we have

$$\left(H^0 + \lambda H^1\right)\left(|n^0\rangle + \lambda|n^1\rangle + \lambda^2|n^2\rangle + \dots\right) = \left(E_n^0 + \lambda E_n^1 + \lambda^2 E_n^2 + \dots\right)\left(|n^0\rangle + \lambda|n^1\rangle + \lambda^2|n^2\rangle + \dots\right).$$

We're now ready to match the two sides term by term in powers of λ.

The zeroth-order term, of course, just gives back $H^0|n^0\rangle = E_n^0|n^0\rangle$.

FIRST-ORDER TERMS

Matching the terms linear in *l* on both sides:

$$H^0|n^1\rangle + H^1|n^0\rangle = E_n^0|n^1\rangle + E_n^1|n^0\rangle.$$

This equation is the key to finding the first-order change in energy E_n^1. Taking the inner product of both sides with $\langle n^0|$:

$$\langle n^0|H^0|n^1\rangle + \langle n^0|H^1|n^0\rangle = \langle n^0|E_n^0|n^1\rangle + \langle n^0|E_n^1|n^0\rangle,$$

then using $\langle n^0 | H^0 = \langle n^0 | E_n^0$, and $\langle n^0 | n^0 \rangle = 1$, we find

$$E_n^1 = \langle n^0 | H^1 | n^0 \rangle.$$

Taking now $\lambda = 1$, we have established that the first-order change in the energy of a state resulting from adding a perturbing term to the Hamiltonian is just the expectation value of H^1 in that state.

For example, we can estimate the ground state energy of the helium atom by treating the electrostatic repulsion between the electrons as a perturbation. The zeroth-order ground state has the two (opposite spin) electrons in the ground state hydrogen-atom wave function (scaled for the doubling of nuclear charge). The first-order energy correction E_0^1 is then given by computing the expectation value $\left\langle e^2 / r_{12} \right\rangle$ for this ground state wave function.

The general expression for the first-order change in the *wave function* is found by taking the inner product of the first-order equation with the bra $\left\langle m^0 \right|, m \neq n$,

$$\langle m^0 | H^0 | n^1 \rangle + \langle m^0 | H^1 | n^0 \rangle = \langle m^0 | E_n^0 | n^1 \rangle + \langle m^0 | E_n^1 | n^0 \rangle.$$

The last term is zero, since $\langle m_0 | n_0 \rangle = 0$, and in the first term $\langle m^0 | H^0 = \langle m^0 | E_m^0$, so

$$\langle m^0 | n^1 \rangle = \frac{\langle m^0 | H^1 | n^0 \rangle}{E_n^0 - E_m^0}$$

and therefore the wave function correct to first order is:

$$|n\rangle = |n^0\rangle + |n^1\rangle = |n^0\rangle + \sum_{m \neq n} \frac{|m^0\rangle \langle m^0 | H^1 | n^0 \rangle}{E_n^0 - E_m^0}.$$

THE SECOND-ORDER ENERGY TERM

To find the *second*-order correction to the energy, it is necessary to match the second-order terms in

$$\left(H^0+\lambda H^1\right)\left(\left|n^0\right\rangle+\lambda\left|n^1\right\rangle+\lambda^2\left|n^2\right\rangle+\ldots\right)=$$
$$\left(E_n^0+\lambda E_n^1+\lambda^2 E_n^2+\ldots\right)\left(\left|n^0\right\rangle+\lambda\left|n^1\right\rangle+\lambda^2\left|n^2\right\rangle+\ldots\right)$$

giving:

$$H^0\left|n^2\right\rangle+H^1\left|n^1\right\rangle=E_n^0\left|n^2\right\rangle+E_n^1\left|n^1\right\rangle+E_n^2\left|n^0\right\rangle.$$

Taking the inner product with $\left\langle n^0\right|$ yields:

$$\left\langle n^0\right|H^0\left|n^2\right\rangle+\left\langle n^0\right|H^1\left|n^1\right\rangle=E_n^0\left\langle n^0 \middle| n^2\right\rangle+E_n^1\left\langle n^0 \middle| n^1\right\rangle+E_n^2\left\langle n^0 \middle| n^0\right\rangle.$$

The leading terms on the two sides cancel as before. What about the term $E_n^1\left\langle n^0 \mid n^1\right\rangle$? Since $|n\rangle=\left|n^0\right\rangle+\left|n^1\right\rangle$, and both $|n\rangle, \left|n^0\right\rangle$ are normalized, $\left\langle n^0 \middle| n^1\right\rangle+\left\langle n^1 \middle| n^0\right\rangle=0$ in leading order that is to say, $\left\langle n^0 \middle| n^1\right\rangle$ is pure imaginary. That just means that if to this order $|n>$ has a component parallel to $|n^0>$, that component has a small pure imaginary amplitude, and $|n>$ can be written (to this order) as $|n\rangle=e^{i\alpha}\left|n^0\right\rangle+$ kets $\perp \left|n^0\right\rangle$, with α small. But the phase factor can be eliminated by redefining the phase of $|n>$, and with that redefinition $|n1>$ has *no* component in the $|n^0>$ direction, we can therefore drop the term $E_n^1\left\langle n^0 \mid n^1\right\rangle$.

So the second-order correction to the energy is:

$$E_n^2=\left\langle n^0\right|H^1\left|n^1\right\rangle=\left\langle n^0\right|H^1\sum_{m\neq n}\frac{\left|m^0\right\rangle\left\langle m^0\right|H^1\left|n^0\right\rangle}{E_n^0-E_m^0}=\sum_{m\neq n}\frac{\left|\left\langle m^0\right|H^1\left|n^0\right\rangle\right|^2}{E_n^0-E_m^0}.$$

SELECTION RULES

Perturbation theory involves evaluating matrix elements of operators. Very often, many of the matrix elements in a sum are zero obvious tests are parity and the Wigner-Eckart theorem. These are examples of *selection rules*: tests to find if a matrix element may be nonzero.

THE QUADRATIC STARK EFFECT

When a hydrogen atom in its ground state is placed in an electric field, the electron cloud and the proton are pulled different ways, an electric dipole forms, and the overall energy is lowered.

The perturbing Hamiltonian from the electric field is $H' = e\varepsilon z = e\varepsilon r\cos\theta$, where ε is the electric field strength, the field is in the z-direction, the electron charge *e* is negative.We shall denote the unperturbed eigenenergies of the hydrogen atom by $E_n = E_{nlm} = -1/n^2$, so in particular we denote the ground state energy by E_1.

The first-order correction to the ground state energy $E_1^1 = \langle 100 | e\varepsilon z | 100\rangle$, where

$$|100\rangle \equiv \psi_{100}(r) = \left(\frac{1}{\pi a_0^3}\right)^{1/2} e^{-r/a_0}.$$

This first-order term is zero since there are equal contributions from positive and negative z.

The second-order term is

$$E_1^2 = \sum_{n\neq 1,l,m} \frac{|\langle nlm | e\mathcal{E}z | 100\rangle|^2}{E_1 - E_n}$$

where we are now using $|nlm>$ to denote the unperturbed hydrogen atom wave functions, and here the $E_n = -1/n^2$ (in Rydbergs) are the unperturbed energies.

Most of the terms in this infinite series are zero the selection rules help get rid of them as follows: since $e\varepsilon z$ is the m = 0 component of a spherical vector and $|100>$ is a zero angular momentum state, it follows from the Wigner-Eckart theorem that. $\langle nlm|$ can only be $\langle n10|$ This reduces the second-order sum over states to:

$$E_1^2 = \sum_{n\neq 1} \frac{|\langle n10 | e\mathcal{E}z | 100\rangle|^2}{E_1 - E_n}.$$

This is still not easy to evaluate, but an *upper bound* can be found be observing that $|E_1 - E_n| \geq |E_1 - E_2|$, so

$$|E_0^2| < \frac{1}{E_2 - E_1} \sum_{n=1} |\langle n10|e\mathcal{E}z|100\rangle|^2 = \frac{1}{E_2 - E_1} \sum_{n,l,m} \langle 100|e\mathcal{E}z|nlm\rangle\langle nlm|e\mathcal{E}z|100\rangle$$

where we have temporarily *restored* the full sum over *n, l, m*, that is, we've put back all the zero terms. The reason for this seeming backward step is that, having taken the energy-difference denominator outside the sum, we can even include $|100\rangle$ in the $|n/m\rangle$ sum (it's another zero term) and in fact we can even include the plane-wave (ionized) states as well as the bound states, since the plane waves all have energy greater than zero. At this point, the $\sum_{n,l,m}$ sum becomes a sum over all states, and therefore just becomes the unit operator,

$$\sum_{n,l,m} |nlm\rangle\langle nlm| = I,$$

so $|E_1^2| < \dfrac{1}{E_2 - E_1} \langle 100|(e\mathcal{E}z)^2|100\rangle.$

For the ground state hydrogen wave function, $\langle 100|z^2|100\rangle = a_0^2$, $E_1 = -e^2/2a_0$, $E_2 = E_1/4$ so

$$|E_1^2| < \frac{1}{\left(\frac{3}{4}e^2/2a_0\right)} (e\mathcal{E})^2 a_0^2 = \frac{8}{3}\mathcal{E}^2 a_0^3.$$

Furthermore, since all the terms in the series for E_1^2 are negative, *the first term sets a lower bound on* $|E_1^2|$:

$$|E_1^2| > \frac{|\langle 210|e\mathcal{E}z|100\rangle|^2}{E_1 - E_2}.$$

This can be evaluated in straightforward fashion to find $|E_1^2| > 0.55 \times \frac{8}{3}\mathcal{E}^2 a_0^3.$

So, even though we have not actually evaluated the second-order correction to the energy explicitly, we have it bracketed between two values, the lower one being more than half the upper one. Other ingenious methods have been developed to find that the true answer is $|E_1^2| = \frac{9}{4}\mathcal{E}^2 a_0^3$, but in fact the whole problem can be solved exactly using parabolic coordinates.

DEGENERATE PERTURBATION THEORY: DISTORTED 2-D HARMONIC OSCILLATOR

The above analysis works fine as long as the successive terms in the perturbation theory form a convergent series. A necessary condition is that the matrix elements of the perturbing Hamiltonian must be smaller than the corresponding energy level differences of the original Hamiltonian. If H^0 has different states with the same energy, in other words degenerate energy levels, and the perturbation has nonzero matrix elements *between these degenerate levels,* then obviously the theory breaks down. To see just how it breaks down, and how to fix it, we consider the two-dimensional simple harmonic oscillator:

$$H^0 = \frac{p_x^2 + p_y^2}{2m} + \frac{1}{2}m\omega^2\left(x^2 + y^2\right).$$

Recall that for the *one*-dimensional simple harmonic oscillator the ground state wave function is

$$|0\rangle = \left(\frac{m\omega}{\pi\hbar}\right)^{1/4} e^{-m\omega x^2/2\hbar} = \left(\frac{m\omega}{\pi\hbar}\right)^{1/4} e^{-\xi^2/2} \text{ with } \xi = \sqrt{\frac{m\omega}{\hbar}}x, \text{ and}$$

$$|1\rangle = \left(\frac{m\omega}{\pi\hbar}\right)^{1/4} \sqrt{2}\xi e^{-\xi^2/2}.$$

The two-dimensional oscillator is simply a product of two one-dimensional oscillators, so, writing $\eta = \sqrt{\frac{m\omega}{\hbar}}y$, the ground state is $|0\rangle = \left(\frac{m\omega}{\pi\hbar}\right)^{1/2} e^{-(\xi^2+\eta^2)/2}$, and the two (degenerate) next states, energy above the ground state, are

$$|1,0\rangle = \left(\frac{m\omega}{\pi\hbar}\right)^{1/2} \sqrt{2}\xi e^{-(\xi^2+\eta^2)/2}, \quad |0,1\rangle = \left(\frac{m\omega}{\pi\hbar}\right)^{1/2} \sqrt{2}\eta e^{-(\xi^2+\eta^2)/2}.$$

Suppose now we add a small perturbation $H^1 = \alpha m\omega^2 xy$ with α a small parameter. Notice that $\langle 0|H^1|0\rangle = \langle 1,0|H^1|1,0\rangle = \langle 0,1|H^1|0,1\rangle = 0$, so according to naïve perturbation theory, there is *no* first-order correction to the energies of these states. However, on going to second-order in the energy correction, the theory breaks down. The matrix element $\left\langle 1,0\middle|H^1\middle|0,1\right\rangle$ is nonzero, but the two states $|0,1\rangle, |1,0\rangle$ have the same energy! This gives an infinite term in the series for E_n^2.

Yet we know that a small term of this type will not wreck a two-dimensional simple harmonic oscillator, so what is wrong with our approach? It is helpful to plot the original harmonic oscillator potential $\frac{1}{2}m\omega^2\left(x^2 + y^2\right)$ together with the perturbing potential $\alpha m\omega^2 xy$. The first of course has circular symmetry, the second has axes in the directions $x = \pm y$, climbing most steeply from the origin along $x = y$, falling most rapidly in the directions $x = -y$. If we combine the two potentials into a single quadratic form, the original circles of constant potential become ellipses, with their axes aligned along $x = \pm y$.

The problem arises even in the classical two-dimensional oscillator: picture a ball rolling backwards and forwards in a smooth saucer, a circular bowl. Now imagine the saucer is made slightly elliptical. The ball will still roll backwards and forwards through the center if it is released along one of the axes of the ellipse, although with different periods, as the axes differ in steepness. However, if it is released at a point *off* the axes, it will describe a complex path resolvable into components in the two axis directions having different periods.

For the quantum oscillator as for the classical one, as soon as the perturbation is introduced, the eigenkets are in the direction of the new elliptic axes. This is a large change from the original x and y axes, and definitely *not* proportional to the small parameter . But the original unperturbed problem had circular symmetry, and there was no particular reason to choose the x and y axes as we did. If we had instead chosen as our original axes the lines $x = \pm y$, the kets would *not* have undergone large changes on switching on the perturbation.

The resolution of the problem is now clear: *before* switching on the perturbation, *choose a set of basis kets in a degenerate subspace such that the perturbation is diagonal in that subspace.*

In fact, for the simple harmonic oscillator example above, the problem can be solved exactly:

$$\frac{1}{2}m\omega^2\left(x^2+y^2\right)+\alpha m\omega^2 xy=\frac{1}{2}m\omega^2\left[(1+\alpha)\left(\frac{x+y}{\sqrt{2}}\right)^2+(1-\alpha)\left(\frac{x-y}{\sqrt{2}}\right)^2\right]$$

and it is clear that, despite the results of naïve first-order theory, there is indeed *a first order shift* in the energy levels,

$$\hbar\omega \rightarrow \hbar\omega\sqrt{1\pm\alpha} \approx \hbar\omega\left(1\pm\alpha/2\right)$$

THE LINEAR STARK EFFECT

The hydrogen atom, like the two-dimensional harmonic oscillator discussed above, has a nondegenerate ground state but degeneracy in its lowest excited states. Specifically, there are four $n = 2$ states, all having energy -1/4 Ryd :

$$\psi_{200}(r)=\left(\frac{1}{32\pi a_0^3}\right)^{1/2}\left(2-\frac{r}{a_0}\right)e^{-r/2a_0},$$

$$\psi_{210}(r,\theta,\phi)=\left(\frac{1}{32\pi a_0^3}\right)^{1/2}\left(\frac{r}{a_0}\right)e^{-r/2a_0}\cos\theta,$$

$$\psi_{21\pm1}(r,\theta,\phi)=\left(\frac{1}{32\pi a_0^3}\right)^{1/2}\left(\frac{r}{a_0}\right)e^{-r/2a_0}\sin\theta e^{\pm i\phi}.$$

Perturbing this system with an electric field in the z-direction, $H' = e\varepsilon z = e\varepsilon r\cos\theta$, note first that naïve perturbation theory predicts *no* first-order shift in any of these energy levels. However, to second order, there is a nonzero matrix element between two degenerate levels $\langle 200|H'|210\rangle$. All the other matrix elements between these basis kets in the four-dimensional degenerate subspace are zero, so the only diagonalization necessary is within the *two*-dimensional degenerate subspace spanned by $|200\rangle, |210\rangle$, where

$$H^1 = \begin{pmatrix} 0 & \Delta \\ \Delta & 0 \end{pmatrix}$$

with

$$\begin{aligned}\Delta &= \langle 200|H^1|210\rangle \\ &= e\mathcal{E}\left(\frac{1}{32\pi a_0^3}\right)\int_0^\infty\left(2-\frac{r}{a_0}\right)\left(\frac{r\cos\theta}{a_0}\right)^2 e^{-r/a_0} r^2 dr \sin\theta d\theta d\phi \\ &= -3e\mathcal{E}a_0.\end{aligned}$$

Diagonalizing H^1 within this subspace, then, the new basis states are $(|200\rangle \pm |210\rangle)/\sqrt{2}$ with energy shifts $\pm\Delta$, linear in the perturbing electric field.

The states $|21\pm1\rangle$ are not changed by the presence of the field to this approximation, so the complete energy map of the $n = 2$ states in the electric field has two states at the original energy of -1/4Ryd, one state moved up from that energy by Δ, and one down by Δ.

Notice that the new eigenstates $(|200\rangle \pm |210\rangle)/\sqrt{2}$ are *not* eigenstates of the parity operator a sketch of their wave functions reveals that in fact they have nonvanishing electric dipole moment $\vec{\mu}$, indeed this is the reason for the energy shift, $\pm\Delta = \mp 3e\varepsilon a_0 = \mp\vec{\mu}.\vec{\varepsilon}$.

Chapter 26

Electrons in One Dimension

INTRODUCTION: LOOKING FOR SUPERCONDUCTORS AND FINDING INSULATORS

In 1964, Little suggested (Phys. Rev 134, A1416) that it might be possible to synthesize a room temperature superconductor using organic materials in which the electrons traveled along certain kinds of chains, effectively confined to one dimension.

The first satisfactory theory of "ordinary" superconductivity, that of Bardeen, Cooper and Schrieffer (BCS) had appeared a few years earlier, in 1957. The key point was that electrons became bound together in opposite spin pairs, and at sufficiently low temperatures these bound pairs, being boson like, formed a coherent condensate all the pairs had the same total momentum, so all traveled together, a supercurrent. The locking of the electrons into this condensate effectively eliminated the usual single-electron scattering by impurities that degrades ordinary currents in conductors.

But what could bind the electrostatically repelling electrons? The answer turned out to be lattice distortions, as first suggested by Fröhlich in 1950. An electron traveling through the crystal attracts the positive ions, the consequent excess of local positive charge attracts another electron. The strength of this binding, and hence the temperature at which the superconducting transition takes place, depends on the rapidity of the lattice response. This was confirmed by the isotope effect: lattice response time obviously depends on the inertia of the lattice, the BCS theory predicted that for a superconducting element with different isotopic varieties, the ratio of the superconducting transition temperatures for pure isotopes T_2/T_1 was equal to $\sqrt{M_1/M_2}$, M_1, M_2 being the ion masses, the lighter isotope having the higher transition temperature. This was indeed the case.

Little's idea was that the build up of positive charge by a passing electron could be speeded up dramatically if instead of having to move ions, it need only rearrange other electrons. Unfortunately, there were no obvious three-dimensional candidate materials. However, if the conduction electrons moved along a one-dimensional chain, polarizable side chains might be attached, and rearrangement of the electronic charge distribution in these side chains would respond very rapidly to a passing conduction electron, building up a local positive charge. If this worked, order of magnitude arguments suggested possible enhancement of the transition temperature by a factor $\sqrt{M/m}$ over ordinary superconductors, m being the electron mass.

In the 1970's, various organic materials were synthesized and tested, beginning with one called TTF-TCNQ, in which a set of polymer-like long molecules donated electrons to another set, leaving one-dimensional conductors with partially filled bands, seemingly good candidates for superconductivity. Unfortunately, on cooling these materials surprisingly became *insulators* rather than superconductors! This was the first example of a *Peierls transition*, a widespread phenomenon in quasi one-dimensional systems.

The basic mechanism of the Peierls transition can be understood with a simple model. It is a nice example of applied second-order perturbation theory, including the degenerate case. We examine the model and the result below.

It should be added that in some newer materials the Peierls transition is (unexpectedly) suppressed under high pressure, and superconductivity has in fact been observed in organic salts, but so far only at transition temperatures around one Kelvin: Little's dream is not yet realised.

SECOND-ORDER PERTURBATION THEORY

To understand how a one-dimensional conductor might turn into an insulator at low. temperatures, we must first become familiar with the simplest model of a one-dimensional conductor: $H = H^0 + V = \frac{p^2}{2m} + V(x)$

with H^0 a gas of noninteracting electrons on a line, and *V* periodic, that is $V(x+a) = v(x)$,

the potential from a line of ions spaced *a* apart. We'll take the system to have *N* ions in a total length *L*, so

$L = Na$ and to keep the math simple, we'll require periodic boundary conditions.

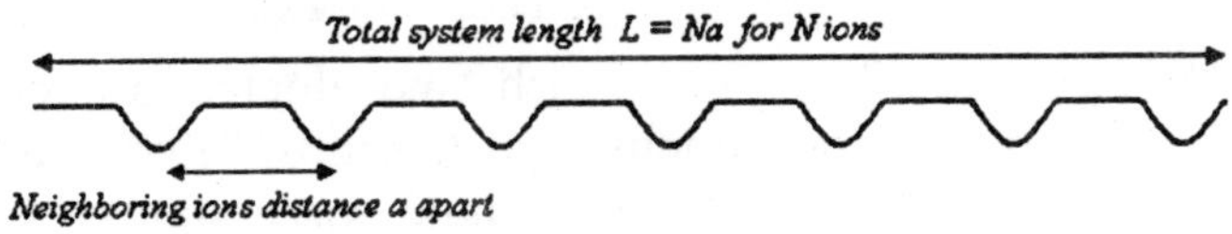

Ionic potential seen by electrons in one-dimensional system.

The physics here is that *without* the potential, the electron eigenstates are *plane waves*. The effect of the lattice potential is to partially reflect the waves, like a diffraction grating, generating components at different wavelengths. This effect becomes particularly important when the electron wavelength matches twice the ion spacing. For that case, the reflected and original waves have the same strength, the electron is at a standstill.

The eigenstates of H^0 are then

$|k\rangle^{(0)} = \frac{1}{\sqrt{L}} e^{ikx}$, with $e^{ikL} = 1$, so $k = \frac{2\pi n}{L}$, *n* being an integer.

The unperturbed energy eigenvalues, $H^0 | k_n\rangle^{(0)} = E_n^0 | k_n\rangle^{(0)}$,

are just $E^0 = \frac{\hbar^2 k^2}{2m}$

(*Note*: this is to be understood as .

$$H^0 \mid k\rangle^{(0)} = E_n^0 \mid k_n\rangle^{(0)}, \text{ with } E_n^0 = \frac{\hbar^2 k^2}{2m} \text{ and } k_n = \frac{2\pi n}{L}$$

We are following standard practice here. We shall also write . $\sum_k f(k)$ meaning $\sum_n f(k_n)$)

It's worth plotting the (E, k) curve:

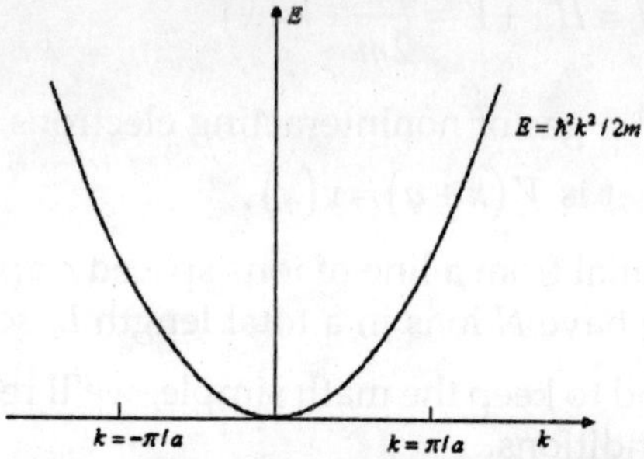

Energy momentum (E, k) curve for a free electron in one dimension.

Suppose we have ions with two electrons each to contribute to this one-dimensional (supposed) conductor. Assuming they move into these plane wave states, in the system ground state they will fill up the lowest energy states up to a maximum k-value denoted by $\pm k_F$ (F stands for Fermi, this is the *Fermi* momentum.) Where is it?

We know there will be a total of $2N$ electrons. We also know that the allowed values of k, from the boundary conditions, are $k_n = 2\pi n / L$, with n an integer. In other words, the allowed k's are uniformly spaced $2\pi n / L$ apart, meaning they have a *density* of $L / 2\pi$ in k-space, so the total number between $\pm k_F$ is Lk_F / π. The $2N$ electrons will have N of each spin, each k-state can take two electrons (one of each spin), so $Lk_F / \pi = N = L / a$, and $k_F = \pi / a$.

To do perturbation theory, we must find the matrix elements of $V(x)$ between eigenstates of H^0:

$${}^{(0)}\langle k' | v | k \rangle^{(0)} = \frac{1}{L} \int e^{i(k-k')} V(x)\, dx$$

This is just the Fourier component v_{k-k} of $V(x)$.If $V(x)$ is periodic with period a,

$V_k \neq 0$ only if $k = nk$, n an integer, k = 2π/a In other words, if a function is periodic with spatial period a, the only nonzero Fourier components are those having the same spatial period a.

Therefore

$$V(x) = \sum_n V_{nK} e^{inKx}, \text{ and } V_{-nK} = V_{nK}^* \text{ since } V(x) \text{ is real; } K = 2\pi / a.$$

The $n = 0$ component of $V(x)$ is of no interest it is just a constant potential, and so can be taken to be zero. Note that this eliminates the trivial first order correction $E_k^1 = {}^{(0)}\langle k|V|k\rangle^{(0)}$ to the energy eigenvalues.

We shall consider *only* the components $n = +1$ and $n = -1$ of $V(x)$, it turns out that the other components can be treated in similar fashion. For $n = +1$, , the potential only has nonzero matrix elements between the plane wave state k and $k + k$, $k - k$ respectively.

So, the second order correction to the energy is:

$$E_k^2 = \sum_{k' \neq k} \frac{\left|{}^{(0)}\langle k|V|k'\rangle^{(0)}\right|^2}{E_k^0 - E_{k'}^0}$$

$$= \frac{\left|{}^{(0)}\langle k|V|k+K\rangle^{(0)}\right|^2}{E_k^0 - E_{k+K}^0} + \frac{\left|{}^{(0)}\langle k|V|k-K\rangle^{(0)}\right|^2}{E_k^0 - E_{k-K}^0}$$

$$= \frac{|V_K|^2}{E_k^0 - E_{k+K}^0} + \frac{|V_{-K}|^2}{E_k^0 - E_{k-K}^0}$$

This result is reasonable provided the terms are small, that is, the energy differences appearing in the denominators are large compared to the relevant Fourier component V_K. However, this cannot always be true! Notice that the state k = π/a has exactly the same unperturbed energy E^0 as the state $k - K = -\pi / a$: in this case, nondegenerate perturbation theory is clearly wrong. In fact, even for states close to , the energy denominator $E_k^0 - E_{k-k}^0$ is small compared with the numerator $|V_{-k}|^2$, so the series is not converging.

QUASI-DEGENERATE PERTURBATION THEORY NEAR THE CRITICAL WAVELENGTH

The good news is that, despite the many states near $k = \pi / a$ and $k = -\pi / a$ that are close together in energy, for any *one* state k near π / a the potential only has a nonzero matrix element to *one* other state close in energy, the state k – K, that is $k = -2\pi / a$. The strategy now is to do what might be called *quasidegenerate* perturbation theory: to diagonalize the full Hamiltonian in the subspace spanned by these two states $|k\rangle^{(0)}, |k-k\rangle^{(0)}$. Other states with nonzero matrix elements to these states are relatively much further away in energy, and can be treated using ordinary perturbation theory.

The matrix elements of the full Hamiltonian in the subspace spanned by these two states are:

$$\begin{vmatrix} E_k^0 & V_K^* \\ V_K & E_{k-K}^0 \end{vmatrix}.$$

Diagonalizing *within this subspace* gives energy eigenvalues:

$$E_{\pm} = \frac{1}{2}(E_k^0 + E_{k-K}^0) \pm \sqrt{\left(\frac{E_k^0 - E_{k-K}^0}{2}\right)^2 + |V_K|^2}.$$

Notice that, provided $\left|E_k^0 - E_{k-k}^0\right| >> |V_k|$, to leading order this gives back $E_{\pm} = E_k^0, E_{k-k}^0$ the order depending on k. However, as k approaches π / a, $\left|E_k^0 - E_{k-k}^0\right|$ becomes of order , and the energies deviate from the unperturbed values. If k is approaching from below, $E_k = E\_ < E_k^0$, and the lower energy is pushed *downwards* by the perturbation: This is a common occurrence with almost degenerate states, perturbations cause the energy levels to "repel" each other.

For k $=\pi/a, E^0_{k-k} = E^0_{k-2\pi/a} = E^0_k$ At this value of *k*, the unperturbed states are exactly degenerate, and the perturbation lifts the degeneracy to give $E_{\pm} = E^0_{\pi/a} \pm |V_k|$.

In the graph below, the green (continuous) curve is the unperturbed energy as a function of *k*, the red curve (with the step) the calculated energy including the leading correction from the periodic potential.

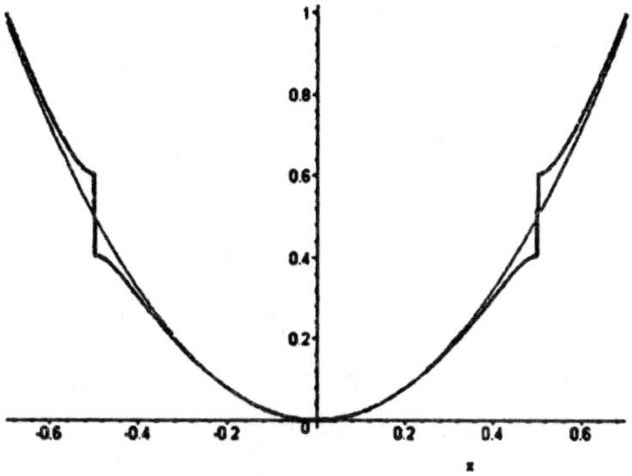

ENERGY GAPS AND BANDS

The energy jump, or gap, of $2|V_K|$ at $|k| = p/a$ means that there are no plane wave type eigenstates with energies in that range attempting to integrate Schrödinger's equation in the periodic potential for such an energy gives exponentially growing and decaying solutions. Such energy gaps in fact are present in real crystalline solids, the allowed energies are said to be in "bands". The lowest band for our model is from $k = -\pi/a$ to π/a Since the allowed values of *k* are given by $k = 2\pi n/L$, the spacing between adjacent *k*'s is $2\pi/L$ and the total number of *k*'s in the lowest band is *L*/*a* = *N*, the same as the number of atoms. Since each electron has two spin states, this implies that a one-dimensional crystal of *divalent* atoms will *just fill* the lowest band with electrons. Therefore, any outside field can only excite an electron to a different state if an energy of at least $2|V_K|$ is supplied for a small electric field, the filled band of electrons will remain in the ground state, there will be no current. This material is an insulator.

On the other hand, if *monovalent* atoms are used, it is clear that the lowest band is *only half full*, adjacent empty electron states are available. The electrons are free to accelerate if an external field is applied. Barring the unexpected, this one-dimensional crystal would be a metal.

Let us now examine how the periodic potential alters the eigenstates. Ignoring the small corrections from plane waves outside the $|k\rangle^{(0)}, |k-K\rangle^{(0)}$ subspace, the eigenstates to this order have the form

$$|k\rangle = a_k |k\rangle^{(0)} + a_{k-K} |k-K\rangle^{(0)}$$

where

$$\frac{a_{k-K}}{a_k} = \frac{E_- - E_k^0}{V_K^*}$$

from the diagonalization of the 2 × 2 matrix representing the Hamiltonian in the subspace.

As k increases from 0 towards , the plane wave initially proportional to e^{ikx} has a gradually increasing admixture of $e^{i(k-2\pi/a)x}$, until at $k = \pi/a$ the two have equal weight meaning that the eigenfunction is now a standing wave. In fact, there are *two* standing wave solutions at $k = \pi/a$, corresponding to the energies below and above the gap. Taking the atoms to have an attractive potential, the lower energy wave has a probability distribution peaking at the atomic positions. The diffractive scattering that gives a left-moving component to a right moving wave is known as Bragg scattering. It also manifests itself in the *group velocity* of the electronic excitations, $v_{group} = d\omega/dk = (1/\hbar)dE/dk$ An electron injected into a one-dimensional metal would not be a plane wave state, but a wavepacket traveling at the group velocity. It is evident that for an injected electron with mean value of k close to π/a, the electron will move very slowly into the metal. This is to be expected the eigenstates become standing waves as $k \to \pi/a$.

For three-dimensional crystals, the situation is far more complicated, but many of the same ideas are relevant. Electron waves are now diffracted by whole planes of atoms, and the three-dimensional momentum space is divided into Brillouin zones, with planes having an energy gap across them.

THE PEIERLS TRANSITION: HOW COOLING A CONDUCTOR CAN GIVE AN INSULATOR

As mentioned in the Introduction, substances very close to monovalent one-dimensional crystals have been synthesized, and it has been found surprisingly that at low temperatures many of them undergo a transition from metallic to *insulating* behaviour. *What happens is that the atoms in the lattice rearrange slightly, moving from an equally-spaced crystal to one in which the spacing alternates, that is, the atoms form pairs.* This is called *dimerization,* and costs some elastic energy, since for identical atoms the lowest state must be one of equal spacing for any reasonable potential. However, the *electrons* are able to move to a lower energy state by this maneuver.

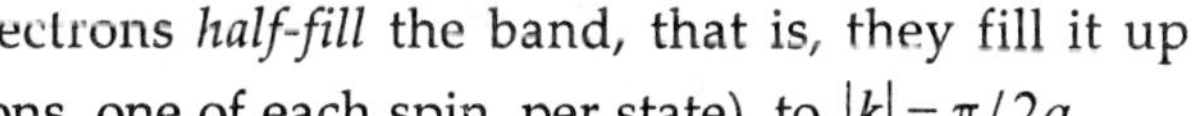

Ionic potential seen by electrons in a dimerized one-dimensional system.

Just how this happens can be understood using the perturbation theory analysis above. For equally spaced atoms, the electrons *half-fill* the band, that is, they fill it up (two electrons, one of each spin, per state) to $|k| = \pi/2a$.

The crucial point is that if the atoms move together slightly into pairs, *the crystal has a new period 2a instead of a.* This means that the *potential* now has a nonzero component at $k = -\pi/a$, with a nonzero matrix element between the states $k = \pi/2a$ and $k = -\pi/2a$, and so on. From this point, we can rerun the analysis above, except that now the gaps open up at $|k| = \pi/2a$ instead of at $|k| = \pi/a$.

The important point is that if the electrons fill all the states to $|k| = \pi/2a$, and none beyond (as would be the case for monovalent atoms) then the opening of a gap at $|k| = \pi/2a$ means that *all the electrons are in states whose energy is lowered.* To find the *total* energy benefit we need to integrate over k.

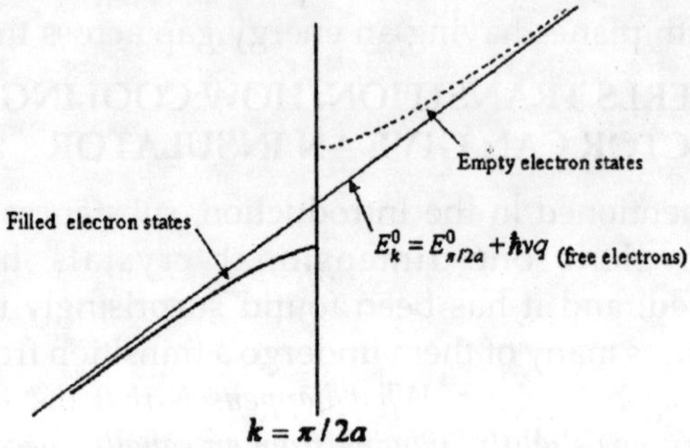

Change in electronic energy levels near $k = \pi/2a$ from dimerization: in this region, the free electron curve is approximated with a straight line: $q = k - \pi/2a$.

CALCULATING THE ELECTRONIC ENERGY GAINED BY DOUBLING THE LATTICE PERIOD

It is evident from the above that most of the contribution comes from fairly close to $k = \pi/2a$ (and of course symmetrically $k = -\pi/2a$). Since we want to find the total lowering in energy, let us study first the *bare* energy as a function of k, that is, the energy with no potential present. Of course, there isn't much to say: $E_k^0 = \hbar^2 k^2/2m$. However, the physics of these one-dimensional systems concerns only excitations near the "Fermi surface", the boundary between filled (low energy) states at zero temperature and empty states. This "Fermi surface" is in fact just two points in one dimension: $k = \pm\pi/2a$. In the neighbourhood of these two Fermi points, it is an excellent approximation to replace the gently curving $E_k^0 = \hbar^2 k^2/2m$ by *straight line approximations* the slope being $dE/dk = \hbar^2 k/m = \hbar p/m = \hbar v$.

Linearizing in the neighbourhood of $k = \pi/2a$, then, we take

$$E_k^0 = E_{\pi/2a}^0 + \hbar v(k - \pi/2a) = E_{\pi/2a}^0 + \hbar vq$$

where

$$q = k - \pi / 2a$$

just k measured from the Fermi point $\pi / 2a$. The variable q is *negative* for the relevant states, since they are on the lower energy side. The density of states in k-space is a constant $2 \times L / 2\pi = L / \pi$, remembering the two spin states per k-value. Recall

$$E_{\pm} = \frac{1}{2}(E_k^0 + E_{k-K}^0) \pm \sqrt{\left(\frac{E_k^0 - E_{k-K}^0}{2}\right)^2 + |V_K|^2}$$

but now $k = \pi / a$, and the lowering of energy of the electrons (counting it as a positive quantity) is:

$$2\int_0^{\pi/2a} (E_k^0 - E_-) L dk / \pi = 2\int \left(\frac{1}{2}(E_k^0 - E_{k-K}^0) + \sqrt{\left(\frac{E_k^0 - E_{k-K}^0}{2}\right)^2 + |V_K|^2} \right) L dk / \pi$$

where the extra factor of 2 counts the symmetrical contribution from the left-hand gap. (In examining the above expression, recall that for the $k > 0$ states we are interested in, $k > \pi / 2a, E_k^0 - E_{k-k}^0$ is *negative*. The integrand on the right-hand side is still positive, very small for small k, reaching a maximum of $|V_k|$ at $k = \pi / 2a$)

Putting in our linearized energy approximation,

$$E_k^0 = E_{\pi/2a}^0 + \hbar v(k - \pi / 2a) = E_{\pi/2a}^0 + \hbar v q,$$

and remembering that now

$$E_{k-K}^0 = E_{-\pi/2a}^0 - \hbar v(k + \pi / 2a) = E_{-\pi/2a}^0 - \hbar v q .$$

Since $E_{\pi/2a}^0 = E_{-\pi/2a}^0$,

$$\left(E_k^0 - E_{k-k}^0\right) = 2v\hbar q$$

Substituting these linearized values in the integral for the total energy lowering:

$$2\int_0^{\pi/2a}\left(E_k^0 - E_-\right)Ldk/\pi = 2\int_{-D}^{0}\left(v\hbar q + \sqrt{(v\hbar q)^2 + |V_K|^2}\right)Ldq/\pi$$

where in terms of the variable q we have set the lower limit of integration at $-D$: we can safely be vague about this lower limit, as the integral turns out to be logarithmic.

Since the integral is over negative numbers, and we have taken the positive square root, it is zero for zero V_K, as it must be.

The integral can be done exactly, but it is more illuminating to divide the range of integration into $|v\hbar q| \le |V_k|$ and $|v\hbar q| > |V_k|$, then estimate the contributions from these two ranges separately.

First, consider $|v\hbar q| \le |V_k|$. Here the integrand is of order $|V_k|$, and the region Δq of integration corresponding to $|v\hbar q| \le |V_k|$ is of order so the integral over this range is of order.

Second, in the region $|v\hbar q| \le |V_k|$, we can write

$$2\int\left(v\hbar q + \sqrt{(v\hbar q)^2 + |V_K|^2}\right)Ldq/\pi = 2\int\left(v\hbar q + |v\hbar q|\sqrt{1+\frac{|V_K|^2}{(v\hbar q)^2}}\right)Ldq/\pi$$

and expand the square root term. The leading terms cancel since q is negative, and the main contribution comes from the next term. This gives:

$$\Delta E \approx 2|V_K|^2\int_{-D}^{-|V_K|}\frac{1}{2\hbar v}\frac{Ldq}{\pi|q|} = \frac{L|V_K|^2}{\hbar v}\ln\frac{|V_K|}{D}.$$

The important thing here is the logarithm. For sufficiently small $|V_k|$, this large (negative) term will dominate any term which is just proportional to V_k^2. But the elastic energy cost of the lattice "dimerizing" the atoms forming pairs, so that the distance between atoms alternates on going along the chain must be proportional to V_k^2. This leads to the conclusion that